Social Protection and Informal Workers in Sub-Saharan Africa

The promotion of social protection in Sub-Saharan Africa happens in a context where informal labour markets constitute the norm, and where most workers live uncertain livelihoods with very limited access to official social protection. The dominant social protection agenda and the associated literature come with an almost exclusive focus on donor and state programmes even if their coverage is limited to small parts of the populations – and in no way stands measure to the needs. In these circumstances, people depend on other means of protection and cushioning against risks and vulnerabilities including different forms of collective self-organizing providing alternative forms of social protection. These informal, bottom-up forms of social protection are at a nascent stage of social protection discussions, and little is known about the extent or models of these informal mechanisms.

This book seeks to fill this gap by focusing on three important sectors of informal work, namely transport, construction, and micro-trade in Kenya and Tanzania. It explores how the global social protection agenda interacts with informal contexts and how it fits with the actual realities of the informal workers. Consequently, the authors examine and compare the social protection models conceptualized and implemented 'from above' by the public authorities in Tanzania and Kenya with social protection mechanisms 'from below' by the informal workers own collective associations.

The book will be of interest to academics in International Development Studies, Political Economy, and African Studies, as well as development practitioners and policy communities.

Lone Riisgaard, PhD, is an associate professor at the Department of Social Sciences and Business at Roskilde University, Denmark.

Winnie Mitullah is a research professor at the Institute for Development Studies, University of Nairobi, with a background in political science and public administration.

Nina Torm, PhD, is postdoc at the Department of Social Sciences and Business at Roskilde University, Denmark.

The Dynamics of Economic Space

This series aims to play a leading international role in the development, promulgation and dissemination of new ideas in economic geography. It has as its goal the development of a strong analytical perspective on the processes, problems and policies associated with the dynamics of local and regional economies as they are incorporated into the globalizing world economy. In recognition of the increasing complexity of the world economy, the Commission's interests include: industrial production; business, professional and financial services, and the broader service economy including e-business; corporations, corporate power, enterprise and entrepreneurship; the changing world of work and intensifying economic interconnectedness.

Agritourism, Wine Tourism, and Craft Beer Tourism
Local Responses to Peripherality Through Tourism Niches
Edited by Maria Giulia Pezzi, Alessandra Faggian, and Neil Reid

Rural-Urban Linkages for Sustainable Development
Edited by Armin Kratzer and Jutta Kister

Beyond Free Market
Social Inclusion and Globalization
Edited by Fayyaz Baqir and Sanni Yaya

Culture, Creativity and Economy
Collaborative Practices, Value Creation and Spaces of Creativity
Edited by Brian J. Hracs, Taylor Brydges, Tina Haisch, Atle Hauge, Johan Jansson and Jenny Sjöholm

Social Protection and Informal Workers in Sub-Saharan Africa
Lived Realities and Associational Experiences from Kenya and Tanzania
Edited by Lone Riisgaard, Winnie Mitullah, and Nina Torm

For more information about this series, please visit: https://www.routledge.com/

Social Protection and Informal Workers in Sub-Saharan Africa

Lived Realities and Associational Experiences from Kenya and Tanzania

Edited by
Lone Riisgaard, Winnie Mitullah, and Nina Torm

Routledge
Taylor & Francis Group
LONDON AND NEW YORK

First published 2022
by Routledge
2 Park Square, Milton Park, Abingdon, Oxon OX14 4RN

and by Routledge
605 Third Avenue, New York, NY 10158

Routledge is an imprint of the Taylor & Francis Group, an informa business

British Library Cataloguing-in-Publication Data
A catalogue record for this book is available from the British Library

Library of Congress Cataloging-in-Publication Data
A catalog record has been requested for this book

ISBN: 978-1-032-00328-3 (hbk)
ISBN: 978-1-032-00329-0 (pbk)
ISBN: 978-1-003-17369-4 (ebk)

DOI: 10.4324/9781003173694

Contents

Illustrations

Figures

Photos

Tables

Boxes

Author Bios

Aloyce Gervas is a PhD student at Mzumbe University, Tanzania under the DANIDA-funded Project 'Informality and Social Protection' that is implemented in Tanzania and Kenya under partnership agreement between Roskilde University Denmark, Mzumbe University Tanzania, and University of Nairobi, Kenya. He holds a bachelor of Public Administration (Human Resource Management) degree from Mzumbe University, Tanzania and a Master of Science in International Management Degreee from the University of Nottingham, United Kingdom. His PhD thesis focuses on Health and Safety Management of Informal Construction Workers in Tanzania.

Raphael Indimuli is a PhD student at the Institute for Development Studies (IDS), University of Nairobi. His PhD focuses on informal workers' participation in Kenya's National Health Insurance scheme. He holds a Master of Arts degree in Development Studies from IDS and a BSc degree in Agriculture and Human Ecology from Egerton University, Kenya. He has recently been involved in a research project conducted in Kenya and Tanzania on social protection and informal workers, funded by the Danish Foreign Ministry. He has also been involved in transport research as a project assistant at IDS under the African Centre of Excellence for Studies in Public and Non-motorised Transport (ACET), a collaborative Research Centre whose focus is on transport issues in African cities.

Anne W. Kamau, DrPH, is a Senior Research Fellow at the Institute for Development Studies (IDS), University of Nairobi, Kenya. She is a medical sociologist with vast research and teaching experience in health and development and social protection in the informal economy. More recently, she has been involved in research on the future of work for women in the transport sector, extending social protection coverage to informal sector workers, and provision of childcare facilities for informal sector working mothers in and around markets. Her recent publications include book chapters focusing on social protection among women small-scale traders in Kenya and the socially just public transport.

Godbertha Kinyondo is a senior lecturer at Mzumbe University, Tanzania, and an economist with over 20 years of experience in research, project evaluation, and teaching. She is conversant in international development and trade issues. Her recent research looks at the evolution of capital mobility in 37 Sub-Saharan African countries.

Winnie Mitullah is a research professor at the Institute for Development Studies, University of Nairobi, with a background in political science and public administration. She is specialized in policy, institutions, and governance, with research focusing on provision and management of urban services. She has researched, taught, consulted, and collaborated with several academic colleagues across countries, regions, and globally. Her 2020 publications include two book chapters: 'Africa and the Global Economy', and 'African Cities and Competitiveness', in Kresl, P.K. (ed.) *Urban Competitiveness in Developing Economies*, Routledge; and 'Gender Mainstreaming and Campaign for Gender Equality', in Nic, C., Karuti, K. and Lynch, G. (eds.), *Oxford Handbook of Kenyan Politics.*

Lone Riisgaard, PhD, is an associate professor in the Department of Social Sciences and Business at Roskilde University, Denmark. She has through her academic career carried out extensive research related to workers' conditions and, in particular, workers' collective agency. She has been involved in research related to the governance of agricultural value chains, regulation through private sustainability standards, and the regulation and agency of labour in global value chains. More recently, she has focused on the collective agency of informal workers in relation to their access to social protection and more broadly with regard to representation.

Nina Torm has a PhD in development economics from Copenhagen University, but is based at the Department of Social Sciences and Business at Roskilde University, Denmark. Her research interests include enterprise and worker dynamics, labour market regulation, and most recently, the interplay between social protection, worker organization, and informality. Nina's work is mostly published in the fields of development economics and labour studies, and geographically, her focus has been on the regions of East Asia (Vietnam and Indonesia) and East Africa (Kenya and Tanzania). In addition to her academic achievements, she regularly consults for different international organizations, including the International Labour Organization and the World Bank.

Acknowledgements

The research in Kenya and Tanzania, which culminated in the writing of this edited book, would not have been possible without the input and support from a number of actors. We highly acknowledge the efforts of the researchers who did not only participate in gathering information but also in analysis and writing which made the publication of this volume successful. However, any errors or omissions remain ours entirely. Our greatest gratitude goes to the informal sector workers in construction, micro-trade, and transport, who took time from their work to answer our research questions, which produced this outcome. We are immensely grateful to the associations and trade unions which shared their experiences with us and all the researchers and assistants who helped collect data in Dar es Salaam, Dodoma, Kisumu, and Nairobi. Their tireless work enabled the gathering of data for this study.

We thank government institutions, agencies, and private-sector organisations in both Kenya and Tanzania that granted us access and participated in Key Informant Interviews and policy dialogues. Appreciation goes to the University of Nairobi (UoN) and, in particular, the Institute for Development Studies (IDS) and Kisumu UoN campus for supporting the Kenya research team; Mzumbe University, and especially, the Dar es Salaam campus for supporting the Tanzanian research team; Roskilde University's department for Business and Society for supporting the research efforts. Finally, this research would not have been possible without the financial support of the Danish Ministry of Foreign Affairs (DANIDA), who generously funded the four-year collaborative research integrating scholarship for two PhD students.

We specifically would like to thank colleagues from Mzumbe University, University of Nairobi, and Roskilde University, who provided insight and expertise that greatly assisted the research. In particular, we are indebted to the late Prof Flora Kessy, and Professor Anna Mdee for their insight and participation in the initiation and early parts of the research. Our deepest appreciation also goes to the Project Assistant Paschalin Basil, who tirelessly oversaw data collection, planning of workshops, and overall management of the Kenyan part of the study. In tandem, we appreciate the efforts of our data management expert Samuel Balongo. Sincere gratitude goes to Peter Gibbon, Jacob Rasmussen, and Peter Lund-Thomsen for insightful comments that greatly improved this manuscript and for fruitful cooperation with the Danish Trade Union Development Agency (DTDA), especially Kent Jensen and Jørgen Assens who generously shared their knowledge and facilitated contacts. We also extend our appreciation to Marty Oehme, who carefully and diligently prepared the manuscript for the publishers.

Finally, we are immensely grateful to our families for their support, motivation, and patience throughout the writing of this book, not least to those no longer with us who inspired and encouraged us to pursue this particular path.

Preface

In the era of structural adjustment, the promise of Universal Social Protection, provided mainly through the state, was formally abandoned by most African governments. Those that retained the promise saw it made empty by the consequent retrenchment in public services – not only in the social sector but also in agriculture, where most economic activity took place at this time.

In the 2010s, when economic growth eventually recovered to levels of the immediate post-independence era, the slogan of Universal Social Protection was once again embraced by African governments. In several cases, its (re-)implementation has even begun, particularly in the form of Universal Health Care.

Paradoxically, this occurred just as economic informalization reached new peaks across the continent. Surveys show that around 80 percent of employment across Sub-Saharan Africa is in the informal economy, driven primarily by de-agrarianization. Since most systems of Universal Social Protection are funded through taxes or other contributions collected via employers, and since a large majority of the African workforce lack a formal employer, this raises questions about financial viability. At the same time, where 'self-contributory' options to obtain coverage for Universal Health Care are offered, the lumpy and precarious character of most informal economy incomes means that take-up amongst their recipients is typically low.

On the other hand, a significant minority of participants in the informal economy are able to access one form or another of non-state social protection. Mostly, this takes place through collective organizations based on the informal sector itself. This is the subject of this important book, which by showing how they work, their benefits and their limitations, illuminates how Universal Social Protection might get nearer universality in reality and not merely in name.

Key to understanding this is to abandon for good the 'heroic' narrative of informality initiated in the 1990s by figures like Hernando do Soto and Naomi Chazan, which saw it as a bastion of rugged individualism and as the cornerstone of a radically anti-collectivist civil society. As the research here shows, between a third and a half of informal workers surveyed belong to organizations anchored in and shaped by their branch of economic activity. Not only is provision of short-term social protection according to the rhythm of employment and unemployment in specific informal branches the main aim of many of these organizations, but in a number of cases, they also provide a bridge between members' and public self-contributory schemes. This can occur, for example, by channelling members' savings into a schedule of payments for premiums.

For the authors of this work, recognizing the anchorage of informal workers' organizations and forms of social protection provision in the nitty-gritty of specific segments of the informal economy goes hand in hand with mapping the latter's extreme heterogeneity of activities, entry barriers, pay and payment systems, working conditions, and relations with public authority. Unusually, it also allows them to map the primacy or subordination by sector and branch of the two faces of the informal worker – worker and business person. Linked to this are the nature of the non-social protection functions that collective organizations may carry out, for example, whether they provide new business leads or act as informal employment exchanges.

In this way, Riisgaard, Mitullah, Torm, and their collaborators end some distance from their starting point in social policy. However, it is worth noting that their journey shares the same coordinates as that of William Beveridge, the founder of the Anglo-Saxon version of Universal Social Protection. The bedrock for *Social Insurance and Allied Services* (1942) was his earlier rigorous study of what was then called 'manpower' – and its distribution across sectors, occupations, classes, forms of employment, and income levels.

Peter Gibbon, May 2021

Introduction

Lone Riisgaard, Nina Torm and Winnie Mitullah

Social protection policies and instruments have gained increasing attention amongst many developing countries and donors since the late 1980s. This is to such a degree that the social protection agenda is now commonly highlighted by development actors and governments alike as the preferred solution to issues of poverty in the Global South (Deacon, 2007; Hickey & Seekings, 2017; Hickey et al., 2020). In Sub-Saharan Africa (SSA), the last decade has witnessed an expansion of social protection schemes such as cash transfers, health insurance, and pensions. However, apart from a few exceptions (Niño-Zarazúa et al., 2012; Hickey et al., 2020), SSA remains largely absent in the growing body of literature on social protection.

The expansion of formal social protection in SSA happens in a context where informal labour markets constitute the norm and where most workers live uncertain, precarious livelihoods with very limited access to official social protection. Dominant ideas about measures like health insurance and pensions or representation and voice have commonly been based on notions that the attainment of such social protection is linked to institutions and workers associated with formal sector employment (du Toit & Neve, 2014). Given that this has left most of the world's workers (who are found in the informal economy) without formal social protection, recent years have seen an opening space for providing informal workers access to social protection schemes. Nonetheless, for a variety of reasons, most current social protection efforts are still largely beyond the reach of informal workers.

At the same time, a myriad of cases exists of informal workers forming collective initiatives, as a means of gaining better representation or for claiming rights to, or providing alternative forms of, social protection. Examples include informal micro-traders in Tanzania lobbying for inclusion of their collective associations in the national health insurance scheme; daladala[1] driver associations pooling resources to buy vehicles to use in times of unemployment or construction worker associations in Kenya providing welfare, training, and jobs for members.

The dominant social protection agenda and the literature on social protection come with an almost exclusive focus on donor and state programmes. Nonetheless, the coverage of the majority of these formal programmes is limited to small parts of the populations and in no way stands measure to the needs. In these circumstances, people elaborate other means of protection and cushioning against

DOI: 10.4324/9781003173694-1

risks and vulnerabilities through extended family ties and, importantly, through different forms of collective self-organizing. These latter collective, informal, bottom-up forms of social protection are notably absent from social protection discussions (for a rare exception, see Awortwi & Walter-Drop, 2018), and little is known about the extent or the format of these informal social protection mechanisms. It is this soaring gap in policy and literary focus that we start to address in this book. Recognizing the very heterogeneous nature of the informal economy, we also note a gap in analyzing how formal social protection schemes interact with the realities and needs of particular groups of informal workers.

This edited volume presents the findings of a 5-year (2017–2021) research project on informal worker associations and social protection.[2] The project has addressed the mentioned gaps by posing the following overall research question:

How are informal workers in the transport, construction, and micro-trade sectors in Kenya and Tanzania accessing formal and informal social protection and what (if any) is the role of their collective associations in facilitating this?

This overall question is specified into three more specific sub-questions:

a Do informal workers associations offer any kind of informal social protection, and if so, what characterizes the format of these services, who benefits from them, and how do they compare to formal social protection measures?

b In the formal social protection schemes, how are social protection needs, delivery, and beneficiaries qualified and to what extent do they cater for informal workers? What, if any, is the role of informal associations in providing access to formal social protection schemes?

c To which extent is representation of informal workers' viewpoints and realities institutionalized or promoted?

Overall, this book sets out to investigate how the global social protection agenda actually touches down in particular contexts and how it relates to the realities of informal workers. In examining and comparing the social protection models conceptualized and implemented 'from above' by the public authorities in Tanzania and Kenya with what people actually do on the ground, we uncover the challenges faced by particular marginalized groups (urban informal workers in three sectors prone to informality) and how their own collective associations seek to address these challenges with what could be called social protection mechanisms 'from below'. With this comparison, we contribute analytically to essential ongoing discussions of social protection models in a Global South informality context.

The research presented in this book is situated in the intersection between the literatures on social protection, on informality, and on collective organizing. Discussions within these bodies of literature have, however (apart from a few partial exceptions like Lindell (2010); du Toit & Neve (2014); Awortwi & Walter-Drop (2018); Alfers & Moussié (2019), largely been kept separate. As such, we seek to instigate conversations between these literatures in order to analytically address empirical issues, which remain largely unexplored in the academic literature.

With regard to social protection, we adopt a broad understanding and more specifically we draw on the typology of social protection elaborated by Devereux and Sabates-Wheeler (2004) and distinguish between social protection classified as protective, preventive, promotive, and transformative measures. Preventive measures encompass not only formal social insurance programmes such as pensions, health insurance, or maternity leave but also informal insurance and cushioning mechanisms. Promotive measures include, for example, access to micro-finance and vocational training, while transformative measures are included primarily in the form of representation and voice. We do not, however, address protective measures such as narrowly targeted safety net measures (commonly known as social assistance), since that falls outside the common assistance of most informal associations. The typology allows us not only to order our empirical data but also to highlight how some social protection measures predominate over others.

From the informality literature, we draw on approaches that emphasize the deeply political implications of informality perspectives. Hence, in this book, we explore not only the actual practices, strategies, and struggles of informal workers (and their associations), but also how informality is understood and governed in Kenya and Tanzania, and in continuation, how informal workers are perceived and governed. This is important not only to understand the realities of informal lives but also for understanding social protection models, as informality perspectives influence how actors in the informal economy are envisioned to access social protection and what type of social protection is seen as relevant for informal versus formal workers.

Investigations into how informal workers are perceived as well as explorations of the actual practices, strategies, and struggles of informal workers and their associations necessarily touch upon the question of agency and what kind of collective capacities informal workers hold and are perceived to hold. Analytical approaches to analyzing the collective capacity of informal workers remain sparse, but, in this volume, we apply and adapt the power resources approach (PRA) (Wright, 2000; Silver, 2003; Schmalz, Ludwig & Webster, 2018). While the approach was developed to analyze the power resources available to trade unions, we argue that, if modified, it still has some explanatory power, particularly in highlighting sectoral differences and especially with regard to issues of transformative social protection in the form of representation and voice.

A key contribution of this book is thus to combine relevant insights and adopted conceptual frameworks from three poorly connected literatures. We need this in order to explore how informal workers in different sectors access both formal and informal social protection in the light of dominant perspectives on informality and social protection and keeping at the centre of attention the role played by informal workers' own collective associations. In doing so, another key contribution of this book is the unravelling of the nexus between formal and informal social protection.

Empirically and methodologically, we contribute to filling a gap in the literature by focusing on three important sectors of informal work, namely transport, construction, and micro-trade in Kenya and Tanzania. The study hence takes a comparative approach exploring three different sectors in two countries with different historical development orientations.

Based on our findings, we argue that there is a need to conceptually rethink and broaden both academic and policy discussions on social protection in order to recognize and address the restrictive formal/informal dichotomy and one-sided focus on formalization as this renders most existing formal social protection measures starkly inappropriate for the majority of the working populations.

The immense importance of informal workers' collective agency in meeting (even if inadequately) the social protection needs of association members needs to not only be recognized but also help inform efforts to reframe national social protection policies and systems. Of particular importance is the representational deficit we identify. Representation should be included in social protection discourses as it is of key importance in terms of ensuring that informal workers have a say in the elaboration of social protection policies and more generally in issues affecting their work and living conditions.

This book is an edited volume where each chapter is authored by different constellations of researchers which all formed part of the research team and the chapters follow the same overall research design, although adapted slightly to fit sectoral and country differences (as discussed in more detail in the section "Book Contributions" at the end of this introduction).

In this introductory chapter, we start by reviewing how social protection, informality, and collective organizing in the informal economy have been approached in the literature while at the same time identifying and elaborating our conceptual framework. Thereafter, we provide a brief overview of the project methodology including a snapshot of the informality phenomenon in our two case countries. Finally, we introduce the chapter contributions to this edited book.

Approaches to social protection

As mentioned, social protection policies and instruments have, since the late 1980s, gained increasing attention amongst both donors and countries in the Global South where many countries have formulated national social protection policies and embraced different forms of social protection (Deacon, 2007; Hickey & Seekings, 2017; Awortwi & Walter-Drop, 2018; Hickey et al., 2020). In the 1990s, social protection instruments were seen as the preferred means to mitigate the adverse impact of structural adjustment (Adésínà, 2011, p. 454) and in the 2000s, following the Millennium Development Goals, the rise of both the post-Washington Consensus and some large Southern powers and the global economic crisis, social protection has been heavily pushed by international development agencies as a tool for poverty reduction (Deacon, 2007; Devereux & Sabates-Wheeler, 2007; Hickey & Seekings, 2017).

Several international and regional policies and recommendations have solidified this policy agenda such as the African Union's 2009 Social Policy Framework, the UN's Social Protection Floor Initiative from 2009, and the International Labour Organization (ILO)'s Recommendation 202 on National Social Protection Floors launched in 2012. Along with the Sustainable Development Goals, which include targets on social protection, these add to the pressure on governments to expand provision (Lavers & Hickey, 2020).

The increased focus on social protection and the increased (although highly uneven) expansion in the Global South of particularly social assistance (in the form of cash-transfers), but also social protection more broadly, has been interpreted in different ways (Lavers & Hickey, 2020). Hanlon, Barrientos, and Hulme (2010), for example, highlight the increased use of social assistance as South-South inspired, following innovative schemes implemented in Brazil, Mexico, and India during the 1990s. Others highlight the importance of international communities such as international organizations, donors, and NGOs in formulating the global social protection agenda and encouraging adoption in the global South (Deacon, 2013; Cherrier, 2016; Simpson, 2017; Ouma & Adésínà, 2018; Hickey et al., 2020). Finally, some authors (e.g. Seekings, 2012; Van de Walle, 2014; Brooks, 2015; Hickey et al., 2020) add to this perspective the importance of local political factors – in particular, democratization processes – in determining how social protection is implemented on the ground. Hickey et al. (2020), for example, focus on social assistance, which they conceptualize as a resource that plays into broader political strategies employed by ruling elites to maintain power. Looking particularly at SSA, they note how the variegated adoption of social assistance needs to be seen in relation to processes of de-agrarianization along with uneven and contested processes of democratization leading to growing political competition but also sometimes to more authoritarian forms of governance.

Despite the current near hegemonic status of the social protection agenda, different understandings of what social protection is and should be flourished, along with a few critical voices denouncing the agenda altogether (e.g. Adésínà, 2011). As noted by Devereux and Sabates-Wheeler (2007, p. 1), the social protection agenda "comes with a fresh array of conceptual frameworks, analytical tools, empirical evidence, national policy processes, heavyweight agencies and big names in development studies aligned behind it". Different agencies push particular understandings of social protection and, thus, for donor-dependent countries, adoption of social protection policies has become somewhat of an imperative alongside the continued emphasis on neoliberal economic policies (Adésínà, 2011). Hence, increasingly, governments in the developing world are elaborating and adopting national social protection strategies as a key feature of their poverty reduction strategies (Barrientos, 2010). Although heavily imposed from the outside, how social protection models have been embraced and adapted in particular national contexts needs to be seen in the context of how these align with the worldview and agendas of local elites (Hickey & Seekings, 2017; Jacob & Pedersen, 2018). In addition, as noted by Kramon (2019), actual implementation also needs to be seen in the context of bottom-up pressures and social norms.

In practice, much literature distinguishes between social insurance and social assistance, while conceptualizations like the rights approach of the ILO also include labour market regulation (Barrientos, 2010). In short, social insurance consists of protection against life contingencies such as maternity, old age, unemployment, or sickness. Social assistance provides support for those in poverty, for example, through conditional and unconditional cash transfers, while labour market regulation refers to labour standards, including rights to organization and representation.

Trajectories on social protection can, however, also be viewed as falling on a continuum ranging from risk management to promoting social justice (Barrientos, 2010; Adésínà, 2011; Gentilini & Omamo, 2011). At one end of the continuum are understandings of social protection as safety nets, as in the original World Bank framework or in its current social risk management approach. Along the continuum, we find broader understandings of social protection like the one promoted by the UK Department for International Development (DfID) which focuses on 'social transfers' and sees social protection from a poverty perspective. Here, we also find the ILO focus on social protection as rights (see Barrientos, 2010; Hickey & Seekings, 2017). Both the ILO and DfID understandings can be characterized as being more along the lines of a European social model of universal provision as opposed to the more minimal approach endorsed by the World Bank (Deacon, 2007; Hickey & Seekings, 2017).

Towards the other end of the continuum, one can place frameworks which incorporate more transformative aims such as the 'transformative social protection' framework developed by Devereux and Sabates-Wheeler (2004) and the UN 'basic needs' approach. Both of these extend the role of social protection to encompass issues like education or representation (which can still be seen as working within a European social model). Although all the trajectories mentioned focus on extreme poverty and vulnerability, the more we move along the continuum, the more emphasis is placed on the broader developmental role of social protection in developing countries, whereas moving left on the continuum, we see increasing trust in the ability of the market to deliver broad economic (and as a result also social) progress.

Placing themselves outside (or beyond) the continuum are a few voices which criticize the current social protection agenda for "(i) excessive focus on the ultra-poor, (ii) the preference for means testing and targeting, and (iii) the disconnection between the 'social' and the broader economic aspects of development policy making" (Adésínà, 2011, p. 459). Particularly, the latter point finds resonance with critique of prevailing discourses on informality and on formalizing the informal as having a narrow focus and refusing "to engage the neoliberal economic framework as vulnerability-inducing" (Adésínà, 2011, p. 458; see also Bernards, 2017). Instead of a narrow focus on social protection, Adésínà (2011) advocates for a transformative social policy approach where attention is given to issues such as universal and publicly funded education, healthcare, and production support for farmers (see also Mkandawire, 2007). Central to advocates of moving 'beyond' social protection is the need to link social and economic policy, to recognize structural conditions of poverty and vulnerability and a call for policies based on norms of equity and solidarity.

From a governmentality perspective, authors (Li, 2007; Hickey & Seekings, 2017) have shown how international development actors have rendered social protection a technical solution to poverty and development challenges more broadly, while sustaining a particular world order based largely on post-Washington Consensus liberal ideas. The influential role that particular international organizations have played in shaping the global policy agenda on social protection in their image has led to what some refer to as the production of "a fragmented and incomplete universalism" (von Gliszczynski & Leisering, 2016, p. 325). The result, as argued by Hickey and Seekings, "is a strategy for promoting poverty reduction that responds to failures in market access and poor governance rather than an effort to fundamentally redistribute wealth and power through a new southern-based model of welfare capitalism" (Hickey & Seekings, 2017, p. 23). In other words, the 'practically feasible solution' to poverty alleviation has been characterized by an absence of measures that go beyond the ameliorative to embrace more transformative forms of social protection.

But how do these conceptualizations relate to more practical operationalizations? To give examples from each end of the continuum, in a narrow sense, social protection can be perceived as public interventions "to assist individuals, households and communities in better managing income risks" (Holzmann & Jorgensen, 1999, p. 4, cited in Barrientos, 2010, p. 9). Whereas in a broader understanding, social protection can be conceptualized as:

> … all public and private initiatives that provide income or consumption transfers to the poor, protect the vulnerable against livelihood risks, and enhance the social status and rights of the marginalised; with the overall objective of reducing the economic and social vulnerability of poor, vulnerable and marginalised groups.
>
> (Devereux & Sabates-Wheeler, 2004, p. iii)

The broader and more transformative understanding illustrated in the quote above by Devereux and Sabates-Wheeler (2004) came as a critical reaction to what they perceived as a narrow 'economic' and safety net focus of earlier policies. They proposed a new analytical framework moving beyond the focus on 'economic protection', to incorporating a 'transformative' element, which recognizes the need for social equity and social rights, as well as protection against livelihood risks. This framework takes social protection beyond the conventional, narrowly specified social insurance/assistance function to include a broader range of both economic and social protection targets/interventions (formal or informal, public or private) classified as protective, preventive, promotive, and transformative measures. As mentioned, in this edited book, we adopt the above typology to structure our empirical data (see Box 1.1).

Historically, social protection policies and practice have undergone significant changes. Prior to the 1940s, 'support to development' was focused on infrastructure, basic native education, and to a limited extent, health – in accordance with the perceptions, interests, and needs of the colonial powers. Although

Box 1.1 Typology of social protection measures as used in this book

Protective measures provide relief from deprivation: they include narrowly targeted safety net measures commonly known as social assistance. Not included in our research since it falls outside what informal associations are likely to assist with.

Preventive measures seek to avert deprivation: they include social insurance for economically vulnerable groups such as formal systems of pensions, health insurance, maternity benefit, and unemployment benefits. They also include informal mechanisms, such as savings clubs and funeral societies as well as strategies of risk diversification. In our project, preventive measures include health and pension insurance systems and saving clubs/welfare associations (in terms of assistance with life contingencies such as sickness, marriage, and death).

Promotive measures aim to enhance real incomes and capabilities through a range of livelihood-enhancing programmes targeted at households and individuals and have income stabilization at least as one objective. In our project, promotive measures include micro-credit, as well as training (business-related) and saving clubs/welfare societies when they function to smoothen income via small loans or enhancing income by offering capital for business development.

Transformative measures seek to address concerns of social equity and exclusion such as collective action for workers' rights or human rights for minority ethnic groups. Interventions include changes to the regulatory framework as well as sensitization campaigns to transform public attitudes and behaviour. In our project, transformative measures include representation, voice, and advocacy.

Source: Based on the framework by Devereux and Sabates-Wheeler (2004).

during World War II, actors like the ILO had advocated for a broader vision of social protection, the single-minded concern with social insurance for the small number of workers in formal employment remained for decades. However, as a result of growing pressure not to ignore the poorer majority of the population in most developing countries, the 1970s saw the birth of the 'basic needs' development strategy, by the ILO, other UN organizations, and the World Bank.[3]

In an African context, post-independence social protection often took the form of the provision of subsidies for education, health, and agriculture, coupled with emergency services (Oduro, 2010). The structural adjustment programmes of the 1980s removed many of these subsidies and introduced user charges for basic social services in health and education (Oduro, 2010). Recognizing the limits of market fundamentalist growth policies, 'adjustment with a human face' brought to the fore the safety nets agenda and targeting of subsidies to the poorest in order to manage risks and shocks (Oduro, 2010; Gentilini & Omamo, 2011).

However, during the late 1990s with the growing realization of the multi-dimensional nature of poverty, safety nets became increasingly criticized as residualist and paternalistic, and more sophisticated alternatives began to be proposed, based also on policy experiments emerging from the Global South, including the Latin American experience with cash transfers. The World Bank's Social Protection and Labour Strategy in 2001 embraced the funding of experimental *conditional* cash transfer programmes, whilst the ILO launched its 'Global Campaign on Social Security and Coverage for All' (2001), based on a recognition that its long-standing commitment to certain forms of social insurance was of little relevance to poor people. In 2005, both the ILO and the World Bank published their first studies of the cost of social protection programmes in Africa (Kakwani & Subbarao, 2005; Pal et al., 2005), and DfID produced their first position paper on cash transfers (DFID, 2005). As discussed further in Hickey and Seekings (2017), DfID was the primary advocate of the form of social protection that would be 'transferred' to the African continent during the mid-2000s, largely in the form of *unconditional* cash transfers.

Hence, in more recent years, cash transfer (conditional and non-conditional) has received increased attention (Adésínà, 2011, p. 454; see also Barrientos, 2010; Oduro, 2010; Hickey & Seekings, 2017; Kramon, 2019; Hickey et al., 2020), and the last decades have seen a rapid extension of particularly social assistance programmes in developing countries, including in East Africa. Examples of such social assistance programmes include the Cash Transfer for Orphans and Vulnerable Children (CT-OVC) programme in Kenya or the 'Productive Social Safety Nets' (PSSN) programme in Tanzania (Barrientos, 2010; Hickey & Seekings, 2017; Kramon, 2019; Hickey et al., 2020).

Developments in understandings of social protection and social protection instruments have been described by some as a journey from universalism to safety nets (Mkandawire, 2004); however, one has to recognize a somewhat broadening of scope in the social protection policy frameworks developed during the 2000s (e.g. by DfID and the ILO) as opposed to the more narrow and passive role played in the 1990s (Gentilini & Omamo, 2011). Recent years have also seen a return to universalistic understandings where some donors, for example, push for universal health coverage. In practice, universal solutions have been implemented in a few African countries, for example, with regard to health in Rwanda. Kenya has recently introduced a universal old age pension scheme (*Inua Jamii Senior Citizens*) and a Universal Health Coverage pilot program, whilst, in Tanzania, universal health coverage has yet to move beyond being a major policy priority, though Zanzibar has universal old age pension.

As argued by Hickey et al. (2020), SSA remains largely absent in the growing body of literature on social protection. In addition, they note how the majority of research on social protection in Africa is donor-commissioned and hence focuses on technical issues, for example, measuring the impacts of particular programmes, assessing different financing models, or examining different targeting and delivery models. As such, Hickey et al. (2020) offer an important contribution to this literature with their comparative examination of the key political drivers behind the uneven expansion of social assistance in SSA. Along with other scholars (Niño-Zarazúa et al., 2012; Seekings, 2012; Lavers & Hickey, 2020), they have noted some distinctive features of African cases, for example, a historical emphasis on social protection, targeted at a predominantly rural population along with current challenges of de-agrarianization such as declining access to agrarian-based community and family support mechanisms, and in general, a context of severe poverty challenges and labour markets which are predominantly informal.

It is in this context that we need to see how the global social protection agenda actually intersects with the realities of lived lives. Apart from the mentioned gap on social protection in SSA more generally, we also note a gap in the literature in seeking to understand how top-down-formulated social protection models interact with particular groups such as the sector-specific groups of informal workers included in our study. Relatedly, we seek to advance the very limited understanding of how such top-down social protection models compare to the myriad of existing coping/protection mechanisms already employed by groups of informal workers, what in this book is named ‘bottom-up social protection’. Individual components of ‘bottom-up social protection’, such as informal savings and loan groups or mutual support schemes, have been studied quite extensively in the literature, but more holistic studies investigating the range, nature, and access to such informal mechanisms are largely absent and, as mentioned, they very rarely figure in the literature or discussions on social protection. The book by Awortwi and Walter-Drop (2018) presents a noteworthy exception in their recent edited volume where they map and analyze informal forms of social protection from a governance perspective to explore what they conceptualize as “governance in areas of limited statehood”. Our contribution is in many ways complementary to their book which also uses the framework of social protection elaborated by Devereux and Sabates-Wheeler; yet, it also goes beyond it in many ways. Most importantly, our book analyzes social protection models in light of the actual challenges faced by informal workers and in the context of specific informal sectors – which enables a focus on the work dimension. Also important is that we compare responses from informal workers on social protection access *in general* – not just informal forms – and thus, we are able to compare formal and informal models of social protection (although as mentioned we do not address social assistance).[4] In addition, we use the PRA (as discussed below) to discuss the strategies and capacities of informal workers.

As mentioned, our research is situated at the intersection between social protection, informality, and collective organizing. Next, we turn our attention towards the vast literature on informality.

Approaches to informality

Informality is a highly disputed concept whether approached as a theoretical concept, as defining a specific part of a population or labour market or as a policy approach. One can rightly argue that the mere use of the term is problematic as it normatively reimposes a dichotomist North/South hierarchy. Nonetheless, we deliberately utilize the terms 'informal' and 'formal', as understandings of and classifications of what is informal or formal have concrete political implications not least with regard to social protection policies and practices.

In the following, we focus mainly on the politics of informality and the links to different underlying logics about desired developments. Policy discourses about informality can be seen as instruments to address poverty, to ensure regulation and hence inclusion of the working population in the formal economy. While this might be the case, policy discourses on informality and the resulting policies implemented are far from just technical instruments. Inspired by a neo-Gramscian and post-development perspective, these discourses in addition can be seen to serve to either deepen or, conversely, aim to transcend existing forms of domination (Bernards, 2017).

One way to categorize different theoretical approaches to informality is to distinguish between neoliberal and structural traditions (Skinner, 2008). Neo-liberal positions have a celebratory view on informality which is perceived as a process of deregulation from below (amongst others, the de Soto inspired legalist approach would fall within this tradition), while structural positions understand informality as a crisis of capitalist development emphasizing the exploitive nature of relations between the formal and informal economies (Skinner, 2008). Several authors have sought to integrate or reframe these two traditions (e.g. Nattrass, 1987; Lourenco-Lindell, 2002; Chen, 2012; Lindell, 2010) or transcend them altogether (e.g. Guha-Khasnobis, Kanbur & Ostrom, 2007; du Toit & Neves, 2014) highlighting how both perspectives offer some explanatory power but also limited analytical value, as they propose what du Toit and Neves (2014) call totalizing teleological frameworks. Du Toit and Neve (2014) suggest instead that, rather than looking at what is not there (i.e. what we would normatively like to be there), we should look at what people actually do and analyze the political terrain, the struggles and strategies employed "by people on the margins of the global economy" (du Toit & Neves, 2014). While we agree with this analytical strategy, policy discourses remain highly important influencers and often informed at least partly by theoretical perspectives. This becomes evident when looking at the discourses practiced and pushed by influential multilateral institutions and donors and also when we look more closely at how informality is understood and governed in Kenya and Tanzania (see Chapter 2 and Chapters 4–9 of this edited book). Thus, while we take up the glove by looking at the actual practices, strategies, and struggles of the very heterogeneous group called informal workers (and their associations), we also place our contribution within a broadly structuralist perspective in that we emphasize the political implications of informality perspectives and how these play into existing power dynamics.

Several authors highlight the very influential role of what they call residual approaches to informality (Bernards, 2017; Rizzo, 2017), in other words, approaches which treat the informal as residue. According to Bernards (2017), two residual understandings of informality have been particularly influential when it comes to the policies implemented in Tanzania and Kenya, namely the policy discourses pioneered by the ILO and Hernando de Soto/the World Bank, respectively.

The understanding of informality within the ILO has widened over time from a dualist perception of an informal sector consisting of unregulated business to an emphasis on unregulated work in an informal economy and a focus on social protection for all (Bernards, 2017). In 2015, the International Labour Conference adopted Recommendation No. 204: 'The Transition from the Informal to the Formal Economy Recommendation'. This recommendation emphasizes the need to move out of informality in order to promote decent work for all (ILO, 2015). Current discussions at the ILO tend to emphasize the exclusion of workers in the informal economy from the opportunities in the formal economy and from social protection. The solution to this perceived exclusion is, on the one hand, to extend certain formal regulations and labour protections to the informal economy (this includes occupational health and safety, freedom from discrimination, forced labour and child labour, as well as the freedom to organize collectively). On the other hand, coverage of social insurances like, for example, maternity protection, as well as decent working conditions, and a minimum wage is envisioned to follow from the transition to the formal economy (ILO, 2015, p. 25; Bernards, 2017).

Critics have highlighted how this approach ignores power relations and structural constraints (Selwyn, 2014; Bernards, 2017; Rizzo, 2017) and promotes what Selwyn (2014) has characterized as "'elite-led' development solutions in which the poor are framed as passive recipients of enlightened governance by states and major corporations" (Selwyn, cited in Bernards, 2017). One could argue that the ILO approach does encourage active participation, as they recommend that employers' and workers' organizations extend membership to workers and economic units in the informal economy so that these can be represented and participate in social dialogue about the transition (ILO, 2015). Nonetheless, the focus is still on the goal of transitioning to formality (quite a few social protection provisions are tied to this transition) and representation through established tripartite organs and not for example on transforming the tripartite structure to enable people working informally to represent themselves on an equal footing alongside formal enterprises and workers.

The other common global policy approach is heavily influenced by Hernando de Soto and presents a celebratory understanding of informality as a form of poor people's empowerment when faced with bureaucratic overregulation by the state (de Soto, 1989; de Soto, 2001). In this view, appropriate institutions need to be developed to promote the entrepreneurial instincts of the poor so that they can successfully engage in capitalist markets. Policy advice includes micro-enterprise development, particularly through access to credit, skills, and property rights and in general, regulatory reforms which decrease barriers to formalization (de Soto, 1989).

Critics have highlighted how the claim that

> 'the poor already possess the assets they need to make a success of capitalism' (de Soto, 2001, p. 6), sits at odds with evidence from a number of studies showing that it is precisely the lack of any asset other than their own labour that forces the very poor to sell their labour power, in precarious and vulnerable conditions, to asset owners.
>
> (Rizzo, 2017, Chapter 1, p. 12)

Again, this approach lacks attention to structural barriers and, in addition, presents a very optimistic view on the ability of the market to deliver development for all.

As argued by Bernards (2017, p. 1834):

> Residualist conceptions of the 'informal' serve to identify the interests of marginal urban workers with the deepening of markets or the extension of prevailing forms of labour regulation (rather than, say, with autonomous forms of political organization or the fundamental de-commodification of labour). Insofar as they are successfully institutionalized in practice, they thus serve to re-articulate the position of existing 'ruling classes'.

Even though the two policy approaches differ in their view on how to achieve development, they both promote formalization by building on and improving existing institutions. In addition, they either endorse – or at least work within – a global market-driven development discourse. The key difference seems to consist in whether formalization is seen mainly as a matter of extending existing labour regulation or whether it is mainly the responsibility of individual workers or what Kamete (2018, p. 184) has described as the difference between a welfarist and a technicist regulatory legalistic form of integration (Kamete, 2018, cited in Steiler, 2018).

In this book, we explore not only the actual practices, strategies, and struggles of informal workers (and their associations), but we also explore popular discourses on informality and how they influence what type of social protection is seen as relevant for informal versus formal workers. This relates to the question of agency and what kind of collective capacities informal workers hold and are perceived to hold. Thus, in the following section, we briefly explore how the collective capacities of informal workers have been addressed in the literature.

Collective capacities of informal workers

In general, earlier literature on informality has tended to assume that people operating within the informal economy lack agency (Lindell, 2010, p. 1). While structuralist approaches recognize the collective agency of workers, the recognition only extends to the formal economy, ignoring the potential agency of workers in the informal economy and often portraying the marginalized urban poor as victims of capitalist development (Lindell, 2010; Egan, 2014). According to Lindell (2010), this widespread oversight of informal worker agency results from not recognizing informal workers as political actors, with the ability to organize collectively (Lindell, 2010, p. 2). This is in line with a critique of the social protection literature by Alfers and Moussié (2019), which argues that, while it might recognize the agency of people in the informal economy, it is as citizens, not as workers. Neoliberal approaches can likewise be criticized for having a one-sided focus on entrepreneurial agency, or in other words, for not recognizing the agency of non-entrepreneurial actors like wage-workers nor the collective agency of informal actors more generally.

The academic critique has coincided with (and been inspired by) actual developments in collective organizing amongst informal workers, and lately, an increasing number of studies have focused more or less explicitly on the ability of informal workers to organize collectively (Chikarmane & Narayan, 2005; Lindell, 2010; Kinyanjui, 2012; Kumar & Singh, 2018; Medina & Schneider, 2019). Kinyanjui (2012), for example, notes that often women in the informal economy belong to informal groups such as merry-go-rounds or women groups commonly known as 'chamas'. Mitullah (2010) argues that such informal institutions are weak on their own but that some have partnered with transnational organizations in organizing and leveraging resources. Such partnerships have made informal workers visible to development actors, including governments, and, in some cases, enabled inclusion in local and national policy processes, which, in turn, have influenced their working environment and livelihoods. Some of the most prominent transnational examples are WIEGO (Women in the Informal Economy Globalizing and Organizing) and StreetNet International, while, at the national level, a coalition like the National Association of Street Vendors of India (NASVI) stands out in particular for the role it has played in the adoption of a national policy on urban street vendors and a street vendors' act (see also the case of the Self Employed Women's Union in South Africa).

As noted earlier in this introduction, discussions within the literature on informality, social protection, and on collective organizing have (apart from a few partial exceptions like Lindell, 2010; du Toit & Neve, 2014; Alfers & Moussié, 2019) largely been kept separate. Drawing on the discussions above, we argue that the common policy advice of formalization along with the widespread lack of recognition of people in the informal economy as political actors with the ability to organize collectively has implications for how actors in the informal economy are envisioned to access social protection. In short, in most of the policy-oriented literature, work-related social protection (like unemployment benefit, paid maternity leave, severance pay, work-injury compensation, etc.) as well as social protection related to voice and recognition is most often closely associated with formal employment and only becomes available to informal actors once they formalize. Alongside policies of formalization, social protection tools like cash transfers and contributory insurances are being promoted to extend social protection coverage to the informal economy. But while formalization and an opening of social insurance schemes is offered in theory as a way to solve this issue, it is less clear that this works in practice, as we shall illustrate with numerous examples throughout this book.

Meanwhile, as also highlighted by Alfers and Moussié (2019) and Hickey and King (2016), social protection approaches concerned with issues of poverty, development, and social justice are mainly oriented towards social assistance and see the key relationship as being that between the citizen and the state. Here, the considerations around work and the relations between capital and labour are absent (Hickey & King, 2016; Alfers & Moussié, 2019). Initiatives along this trajectory include, for example, participatory governance initiatives and community management committees (Hickey & King, 2016). Here, informal workers might have opportunity to participate – but in the capacity of citizens rather than as workers (Hickey & King, 2016; Alfers & Moussié, 2019).

Existing tripartite structures (where labour in the form of trade unions, employers, and the state negotiate labour and social protection policies), while varying in strength across countries (Carré, Horn & Bonner, 2018), largely represent the institutionalization of past social compromises based on the "golden years of prospering welfare capitalism" (Dörre, Holst & Nachtwey, 2009). Informal workers are most often excluded or have limited representation in such tripartite structures. Recent years, however, have shown signs towards some representation of informal workers as, for example, the Urban Areas and Cities Act of Kenya, which provides for informal workers to be represented in City/Municipal Boards. The Kenya Micro and Small Enterprise Authority also has a slot for informal workers. Nonetheless, there are still very few institutionalized spaces where informal workers can directly negotiate as equal social partners in the design of social protection policies (Alfers & Moussié, 2019) or other issues relevant for their working life.

The lack of representation of informal workers in their own right also becomes noticeable when looking at the increasing number of examples of collaboration between informal workers associations and trade unions. Based on experiences from workshops held as part of the Rights Based Social Protection in Africa project, Alfers and Moussié (2019, p. 5) found that

> organizations of informal workers have found collaborations with formal trade unions to be crucial in accessing tripartite fora, creating space for greater participation in social protection policy making and scheme oversight, and in terms of support for the extension of social protection to informal workers.

This has been the case, for example, in Ghana in the development of a new pension scheme and in Zambia where the Alliance of Zambian Informal Economy

Associations (AZEIA) has worked closely with the Zambian Congress of Trade Unions (ZCTU) on ensuring input into social protection policy processes (Alfers & Moussié, 2019). Interestingly, however, they mention how during these workshops, trade unions, and informal worker organizations expressed different perspectives on what have been the most problematic aspects of such collaborations. While trade unions emphasized difficulties in relation to membership fees and organizing meetings, informal worker organizations focused on the question of equal representation within the tripartite structures where social protection policies are discussed (Alfers & Moussié, 2019).

In this book, we contribute to these emerging discussions by investigating how and to which extent the representation of informal workers' viewpoints and realities are institutionalized in Kenya and Tanzania and how this relates to the collective capacities of informal workers associations in the two countries.

How to conceptualize the collective capacity of informal workers

As mentioned earlier, analytical approaches to analyzing the collective capacity of informal workers remain sparse; yet, one approach which has recently been applied and which we also employ in this book is the PRA. The PRA stems from discussions on trade union renewal and has emerged as an analytical tool – often used in close collaboration between practitioners and academics.[5] The PRA starts from the basic premise that, if organized, labour can successfully defend its interests by collectively mobilizing different power resources (Schmalz, Ludwig & Webster, 2018).

The basic concepts on which the PRA built were created by Erik Olin Wright (2000) and Beverly Silver (2003), and they conceptualized the sources of labour power as deriving from both structural power (derived from workers position in the economic system) and associational power (derived from the formation of collective organizations of workers). In the following decade, conceptual additions have been developed by a range of scholars to form an overlapping analytical tool. Institutional power has been added as deriving from "laws, regulations, procedures, practices and other formal and informal rules that formalize the relationship between trade unions, employers and the state and thus secures rights for workers" (Kumar & Singh, 2018, p. 137). Societal power describes "the latitudes for action arising from viable cooperation contexts with other social groups and organizations, and society's support for trade union demands" (Schmalz, Ludwig & Webster, 2018, p. 122). Societal power arises from two different sources, namely coalitional and discursive. 'Coalitional power' refers to alliances with other social actors such as civil society organizations which can be activated to mobilize or campaign, whereas 'discursive power' describes the power to successfully influence public discourses and public opinion (Kumar & Singh, 2018). Finally, Marslev (2019, p. 18) has recently added what he calls 'regime-disruptive power' as a third form of structural power which derives from "the capacity to cause disruption to, and on that basis extract concessions from, political elites interested in regime-survival".

The PRA was developed to analyze the power resources available to trade unions – not informal workers. As such, there are several difficulties in transferring this typology to informal economy workers.[6] For one, they typically have very low marketplace bargaining power (i.e. low entry barriers and a large pool of unemployed labour) and workplace bargaining power (although this differs between sectors). Nonetheless, as also mentioned by Kumar and Singh (2018, p. 137), despite informal workers' weak structural power, "due to their vast numbers, they have the potential to harness significant associational power". We would add that 'regime-disruptive power' is also highly relevant with regard to informal workers as they are recognized as important 'vote banks' and hence in theory should on that basis be able to extract concessions from political elites particularly during election times.

However, the conceptualization of power resources, as developed from the perspective of trade unions, omits sources of power, which we have found to be of relevance to informal workers' associations, for example, controlling access to operate in particular locations or having personalized ties with more powerful actors. In addition, the conceptualization of power resources also rests on the assumption that the main purpose of collective organizing is representation. However, as we shall see in the contributions to this book, many informal workers form collectives for other purposes – for example, to find work or to form informal bottom-up social cushioning/protection measures. Another factor is the multitude of different work and employment relations in the informal economy, which means that there is not always a 'natural' employer to be targeted or engaged with in negotiations. Also, the lines between who constitute workers versus employers are often blurred and there are often quite distinct power hierarchies between workers within specific sectors (see also Rizzo, 2017). This means that one should not assume automatic solidarity within a collective of workers in a sector, sub-sector, or workplace.

This opens the question of how useful the power resource approach is to cast light on informal worker associations. We argue that if modified and employed with caution, the analytical typology still has some explanatory power, especially in emphasizing how sector- and country-specific groups of informal workers have quite differing power resources available and different potentials which might be leveraged, in particular, with regard to issues of transformative social protection in the form of representation and voice. In this edited volume, we combine the PRA with the preventive, promotive, transformative framework thus adding a broader social protection dimension to the analysis, as summed up in the concluding chapter. Below, we highlight our understanding and use of the PRA concepts which should not be seen in isolation, but as partly overlapping and mutually reinforcing.

Typology of power resources as used in this book

In line with Silver (2003), **structural power** can be divided into **workplace bargaining power** (the power that can be utilized from the strategic location of a certain group of workers in a key sector) and **marketplace power** (relating to the tightness of the labour market). As mentioned, informal workers will, in general, command weak marketplace power, in addition, as many informal workers (like e.g. most micro-traders) are not wage earners but own-account workers, their ability to disrupt production processes are very limited. Nonetheless, strategic disruption abilities can be significant in some sectors and is of key importance for informal transport workers who if coordinated can bring the city to a halt (see e.g. Rizzo, 2017; as well as Chapters 4 and 5 of this volume). For informal construction workers, the disruption potential is limited to specific construction sites, whereas, for informal micro-traders, disruptive power is of less potential – save from the latent ability to physically block certain areas or, for example, storm city hall via demonstrations.

Strategic disruption abilities can, if coordinated and scaled, pose a real threat to regime survival and hence transform into 'regime-disruptive power'. For informal economy workers as an overall group, this is a potentially potent source of power, due to their strategic importance as 'vote-banks' during election times specifically. Lately, we have seen how bodaboda[7] drivers have begun to play important roles in campaigning, for example, in Tanzania where the opposition leader would pose with bodaboda riders in the running up to the local elections in 2019 and bodaboda riders wear campaign slogans on their bikes. Interestingly, bodaboda riders are also used strategically in the protests of other groups, as in Kenya, it is now common practice to hire bodabodas to boost street protests regardless of the topic of protest.

Associational power is derived from the formation of collective organizations of workers. Given the context of informal workers, we include all the different types of collective associations created or engaged in by informal workers to advance their own interests and conditions (see also Riisgaard & Okinda, 2018). We also caution to leave it an open empirical question who the relevant subject of workers' strategies and possible counterpart in negotiations might be. Associational power understood in this broad manner is potentially of immense importance for informal workers, as we shall explore throughout the contributions of this edited book. Other than numbers, the literature on PRA mentions how associational power also rests on infrastructural resources, organizational efficiency, member participation, and internal cohesion (Schmalz, Ludwig & Webster, 2018). These are of importance to informal associations as well, as are connections to people in power for some associations.

Institutional power is derived from laws, regulations, procedures, and practices that regulate the relationship between worker associations and employers/authorities. In the context of informal workers, sources of institutional power could be tapping into established tripartite structures via affiliation with trade unions or employers' associations (at government level) and more informally tapping into the access to authorities that trade unions and employers' associations have (at municipal and ward level). As will be illustrated in the later chapters, this has been the case for some micro-traders' associations in Tanzania and for some transport workers in both Kenya and Tanzania. The case of micro-traders also illustrates how access to trading space – whether illegal, legal or simply tolerated – is sometimes the context on which institutional power depends. In the case of informal transport workers in Kenya (see Chapter 5), we see another form of institutional influence in that the legal framework in place for matatu transport is very influential in regulating the relationship between the transport workers and the owners of the vehicles, although not always in ways intended.

Societal power arises from cooperation with other social groups and society's support for worker demands (coalitional power and discursive power). In the context of informal workers, this relates to coalition-making with other informal associations, with formal worker associations, and with broader civil society organizations. In some examples of this book, we found coalitions with other actors, including NGOs and external donors, to be highly related to the ability of associations to advocate for rights and attempts at influencing the image of informal workers from derogatory to contributory.

In the following, we discuss the methodology adopted in the project and very briefly otline the scope of informality in Kenya and Tanzania.

Methodology

In this book, we adopt a broad definition of the informal economy, including self-employment in informal enterprises (i.e. unregistered business) as well as wage employment in informal jobs (i.e. without a written contract,[8] but possibly working for a formally registered enterprise). This broad definition thus also encompasses agricultural employment (which we do not cover) and self-employment. The broader term 'informal economy' which was endorsed by the International Labour Conference (ILC) in 2002 is now commonly used instead of the older and narrower concept of the informal sector. At the same time, it should be clear from our discussion of different approaches to informality that the concept is at the same time highly politicized and in addition empirically unclear, as will be illustrated throughout this book. Hence, informality and formality often come in degrees on a continuum and the formal and the informal intersects in numerous ways. Informal micro-traders might, for example, sell goods on commission for formal retailers while transport workers might be informal but work in formal vehicles. For the purpose of this book, however, it does make sense to employ (with caution) the term informal because it is a significant delimitation in terms of social protection rights. For a brief outline of the scope of informality in Kenya and Tanzania, see Boxes 1.2 and 1.3.

The research presented in this book is based on a combination of qualitative and quantitative data covering informal workers across the sectors of construction, trade, and transport. As seen in Boxes 1.2 and 1.3, in both Tanzania and Kenya, micro-trade, transport, and construction represent some of the key activities in settings, where informality is the norm and shows no signs of decreasing. These two countries thus provide particularly relevant contexts for studying the access that people in the informal economy have to both formal and informal social protection mechanisms – insights that are essential in light of the ongoing expansion of social protection in both Tanzania and Kenya as well as in SSA more broadly.

Box 1.2 The scope of informality in Kenya

In the case of Kenya, the informal economy was estimated to account for 83.6 percent of total (non-agricultural) employment in 2018, an increase by 5.4 percent compared to 2017. Given that the total employment is 17.8 million, and there is 14.9 million in informal employment, this leaves approximately 2.9 million to the formal part of the economy (KNBS, 2019). Furthermore, the informal economy is growing, accounting for 89.7 percent of new jobs created in 2016 and more than 90 percent in 2018 (out of the 840.6 thousand new jobs created in 2018, the informal sector created 762.1 thousand). Given the proportion of the labour force that works within the informal economy, its estimated contribution to GDP remains relatively low at between 20 percent and 24 percent, depending on the source (Medina & Schneider, 2019). Women make up the majority of workers in the informal economy, with men accounting for the larger share of employment in the formal sector (World Bank, 2016). In terms of the division between rural and urban informal economy employment, this was estimated at 9.6 and 5.3 million, respectively, with Nairobi alone accounting for nearly a quarter of total informal employment (KNBS, 2019). Regarding the different sectors, the figures are relatively stable over time with the wholesale, retail trade, hotels, and restaurants industry accounting for approximately 60 percent of total informal economy employment, the manufacturing industry making up an additional 20 percent, and another 10 percent coming from construction, transport, and communication. The remaining industries in the informal economy are community, social, and personal services (together accounting for 10 percent) (KNBS, 2019).

Box 1.3 The scope of informality in Tanzania

Turning to Tanzania, the share of informal employment in total employment is estimated at 90.6 percent (ILO, 2018) with a higher proportion of females (93.1 percent) than males (88.2 percent) (ILO, 2018). Excluding agriculture, the share of informal employment is around 69.3 percent, again with a significantly higher share of females (74.9 percent) compared with males (64.7 percent) (ILO, 2018). Measured by households, 42.5 percent (out of 10.2 million households in Tanzania mainland) have at least one member engaged in informal business, and this represents a slight increase (by 3.0 percent) from 2006 to 2014. In urban areas, more than a half of households are engaged in informal business, accounting for 64.6 percent in Dar es Salaam (up from 56.5 percent in 2006) and 56.9

percent in other urban areas (NBS, 2014). Informal employment makes up 77.6 percent of total urban employment and 96.9 percent of rural employment (ILO, 2018). Regarding the informal economy's contribution to the economy, this is estimated at 47 percent of GDP in Tanzania (Medina & Schneider, 2019). In terms of the different sectors, wholesale and retail trade are the dominant informal activities accounting for 47.9 percent, whilst the second most important activity is accommodation and food service with 14.5 percent, both of which have a higher share of women (NBS, 2014). Manufacturing comes third, corresponding to 9.8 percent, followed by construction activities, which account for 6.2 percent of employment, and finally, transportation (and storage) at 6 percent, with both construction and transportation being dominated by men (ILO, 2018).

Location-wise, data was sampled from two urban areas in each country: Nairobi and Kisumu in Kenya and Dar es Salaam and Dodoma in Tanzania. In each location, three zones/districts were identified through transport hubs, and the same sites were used across the three sectors.[9] In line with our broad definition of the informal economy, the survey covered wage-workers, own-account workers, and micro-businesses with max two employees at any given point in time. In terms of the qualitative data collection, we carried out a series of semi-structured interviews with leaders and members of informal worker associations, totalling 120 KII and 24 FGDs (approximately 10 KII and 2 FGDs per sector per site).[10] We also spoke to representatives from relevant government authorities, trade unions, and business associations in order to assess the broader economic and institutional setting and its role in terms of access to social protection measures for informal workers.

Regarding the quantitative data, we undertook a survey of informal workers using a combination of purpose-based sampling and random selection. More specifically, the sampling was done so that 25 percent of workers were sampled through the formal (registered) associations identified purposively for the project (associations/SACCOs/trade unions) where the target group were ordinary members and not leaders or members in an official position. The reasoning behind this was to ensure a broad coverage of different types of associations amongst our respondents, whilst being aware of the potential bias introduced by purposive-based sampling. The remaining 75 percent of the workers were sampled by geographical location to ensure a degree of randomness, bearing in mind that it is not possible to ensure a representative sample when there is no clearly defined population of informal workers (and hence the probability of selection cannot be specified). This random sampling was also intended to capture members of smaller, more informal worker associations. Nonetheless, due to the purposive-based sampling described above, we are likely to have an over-representation of registered associations as compared to unregistered ones. The final sample was 1,462 workers, which was well above the initial target of 1,200 workers (600 per country divided equally per sector per site).[11]

In terms of the sectors, for the transport sub-sample, we aimed for an equal division between bodaboda and daladala/matatus, and for the latter, both drivers and conductors were sampled, whilst for the bodaboda, the riders (not passengers) were interviewed. For the micro-traders, the target group were mobile ones, whether on the street, at bus terminals, vacant lots, etc., and also less mobile ones, for example, mama lishe, but excluding traders with permanent structures such as kiosks or stalls in regular designated markets. In addition, the enumerators were instructed not to over-cover some commodities, but rather mix different types of traders and different types of commodities. For construction, the target groups were skilled and unskilled workers (wage-workers and own-account) employed directly by construction/site managers or indirectly via an intermediary such as gang-leaders. As for the actual construction sites, we covered large and medium building sites, and waiting sites (streets), yet excluded individual residential housing sites.

The research is comparative in that it explores three different sectors across two country settings. Tanzania has a socialist-oriented economic history with relatively lower inequality and earlier urbanization compared to Kenya. This might have an effect on both the structure of the informal economy and the informal and formal social protection systems. The comparison between Kenya and Tanzania also allows for exploring informal worker organization and social protection access in two different institutional contexts (regulation concerning informal workers is relatively more centralized in Kenya and recently somewhat more inclusive). By comparing a larger urban area (main city) with a smaller city in each country, we seek to address differences (if any exists) which might be related to the degree of urbanization.

In addition to a cross-sector-country comparison/overview, this edited book contains six in-depth cases on the basis of which we compare: (a) the same sectors situated within different country contexts; (b) three different sectors within each country. The comparison of the three sectors within each country allows for uncovering sector specific differences like, for instance, the potentially high disruption ability of transport workers (structural power) or the existence of established umbrella associations of informal micro-traders such as KENASVIT in Kenya and VIBINDO in Tanzania (associational power). It also allows us to compare very different employment setups, where transport is dominated by wage employment, micro-trade by own-account workers, and construction by a combination of both.

The findings presented in this book are drawn from the data described above, with the combination of qualitative and quantitative methods allowing for both breadth and depth in understanding the collective organization and access to social protection amongst informal workers in both larger and smaller urban settings across the three sectors. As such, the book as a whole provides insights into how people in some of the most common and highly vulnerable informal occupations organize and how (if at all) they access formal or informal types of social protection.

In the following, we briefly introduce the contributory chapters of the book.

Book contributions

Chapter 2. *Formal social protection and informal workers in Kenya and Tanzania: From residual towards universal models?* by Nina Torm, Godbertha Kinyondo, Winnie Mitullah, and Lone Riisgaard

This chapter provides an overview of the relevant formal social protection frameworks in Kenya and Tanzania, focussing on how they relate (or not) to informal workers, including recent advances made in this regard. The chapter also positions the formal social protection schemes in relation to the social protection literature and shows that both countries are undergoing a gradual shift from the mostly *residual* towards the more *universal* end of the social protection spectrum, albeit at their own pace. For instance, whilst universal health coverage has become a major policy priority in Tanzania, progress is further ahead in Kenya where a universal health insurance pilot program was carried out in four counties during 2020. As for pensions, newer models range from being fully government funded to relying on informal worker contributions, the latter limiting the effectiveness of the schemes. Although social protection coverage has increased for the general population in both Kenya and Tanzania, when it comes to informal workers, uptake remains limited due to the factors like the cost burden, inflexible procedures, inadequate benefits, and lack of information on the different social protection options.

Chapter 3. *The relationship between association membership and access to formal social protection: A cross-sector analysis of informal workers in Kenya and Tanzania* by Nina Torm

Using survey data of informal workers across the three different sectors (construction, transport, trade) and four urban locations in Kenya and Tanzania, this chapter provides a mostly quantitative assessment of the extent to which informal worker associations facilitate access to both formal and informal social protection measures. Overall and when controlling for key worker characteristics, the analysis shows that association members are significantly more likely to access formal social insurance. Moreover, across sectors and sites, the provision of loans appears to be the main direct mechanism through which this occurs. Finally, the results reveal a substantial earnings-gap between association members and non-members, albeit with differences by sector and country. In sum, the findings suggest that informal worker associations play important roles both in terms of providing direct cushioning and indirectly through enabling participation in formal social insurance schemes.

Chapters 4–9 are all organized in a similar way; however, each pursue in detail the specificities of each sector in their country context elaborating some of the issues raised in Chapters 1–3. The generic chapter content is as follows: First, a section on the context of how the sector is governed and, relatedly, how informal workers in the specific sector are organizing collectively. Second, based on empirical data, the challenges faced by informal workers and their access to formal social protection mechanisms are explored. Third, how the workers in question are organized is discussed along with the services offered by their associations, the power resources available to them, as well as the access barriers. Fourth, the survey data is used to compare informal workers who are association members to non-members and equally compare workers who enrol in formal social protection schemes to ones who do not enrol.

Chapter 4. *Self-regulating informal transport workers and the quest for social protection in Tanzania* by Godbertha Kinyondo

Chapter 4 explores the emerging collective forms of informal transport workers and their potential for enabling access to social protection measures in Tanzania. Although their structural power is limited by a highly competitive sector, informal transportation workers have formed associations that self-regulate the industry and provide partial preventive and promotive social protection measures such as welfare services, savings groups, access to loans, short-term unemployment insurance, and ad-hoc representation. Although these associations understand the needs, priorities, and contributory capabilities of their members better than formal social protection sector institutions, capacity-building is required to improve their effectiveness.

Chapter 5. *Informal transport worker organizations and social protection provision in Kenya* by Anne W. Kamau

Chapter 5 investigates the role of worker organizations in the provision of social protection for informal transport workers in the Kenyan paratransits and motorcycles sub-sectors. The findings reveal that a key benefit of both the paratransit and motorcycle associations is negotiating on behalf of workers when they get into problems with authorities. Further, despite legal provisions to protect workers in the sector, the legally mandated SACCOs largely serve the interests of owners of vehicles and SACCO management. Hence, the main challenge is not the absence of laws but lack of enforcement and supporting systems.

Chapter 6. *Informal trader associations in Tanzania – providing limited but much-needed informal social protection* by Lone Riisgaard

Chapter 6 reveals that even though the micro-trade sector in Tanzania is poorly coordinated, recently, a few attempts at broader coalitions and cooperation have taken place pointing towards a future strengthening of representational power. Whilst most informal traders lack access to social protection benefits, their own associations provide a range of social protection services ranging from savings and loans functions, cushioning in cases of sickness or death in the family and at times vocational training and joint business activities. These informal social protection systems are based on reciprocity and although they extend only limited coverage, they nonetheless provide services which, for most informal micro-traders, are difficult or impossible to access elsewhere.

Chapter 7. *Access to social protection in Kenya: The role of micro-traders' associations* by Raphael Indimuli

Chapter 7 shows how micro-traders in Kenya access both formal and informal social protection services. The chapter reveals that most micro-traders still lack access to formal social insurance and attributes poor enrolment and retention in formal insurance schemes to socio-economic factors such as low and irregular incomes and service provision factors such as cost of premium and cumbersome procedures. Trader associations are revealed to have developed their own forms of social protection to cushion themselves albeit not adequately in terms of meeting all needs. Furthermore, informal micro-trader associations are represented in the government Micro and Small Enterprises Authority, an aspect which has potential of advancing the status of micro-traders in Kenya.

Chapter 8. *Social protection and informal construction worker organizations in Tanzania: How informal worker organizations strive to provide social insurance to their members* by Aloyce Gervas

Chapter 8 shows that, although not many informal construction workers in Tanzania have joined associations, members appreciate the services offered and acknowledge that associations act as cushioning mechanisms to assist workers in a flexible manner during their times of need. By contrast, formal social insurance schemes seem not to respond to construction workers' specific challenges pointing to the need for a more inclusive approach in addressing the low enrolment rate.

Chapter 9. *Construction workers in Kenya: Straddling with formal and informal social protection models* by Winnie Mitullah

Chapter 9 delves into the construction sector in Kenya exploring how construction workers are organized, how workers' associations contribute to labour agency and social protection, and the role played by both formal and informal institutions in strengthening or undermining workers' organizing for social protection. The findings reveal deficits in formal social protection and how workers' associations and networks fill the deficit, albeit inadequately. Going forward, providing social protection for construction workers calls for a combination of bottom-up efforts of construction workers and top-down public models, nuancing the strategies of the latter with the former.

Chapter 10. *Convergence and divergence of workers' environment, associations, and access to social protection: Sectoral and country comparisons* by Winnie Mitullah, Lone Riisgaard, Nina Torm, Aloyce Gervas, Raphael Indimuli, Anne W. Kamau, and Godbertha Kinyondo

Drawing on the preceding chapters, Chapter 10 provides a comparative analysis of the construction, transport, and trade sectors across Kenya and Tanzania revealing many synergies in respect to the working environment and the importance of associations. The major differences which emerge between the sectors relate to the nature of organization, in which the trade and transport sectors manifest a higher level of organization compared to the construction sector. Also, in terms of power resources, worker challenges, and associational benefits, most of the differentiation manifests itself on a sectoral basis, underscoring the importance of adopting a sector-specific lens when analyzing the realities of informal workers.

Chapter 11. *Concluding reflections* by Lone Riisgaard, Winnie Mitullah, and Nina Torm

The concluding chapter offers overall reflections on the research questions and highlights our contributions to the literature and ongoing discussions around social protection and informal workers. The chapter starts by teasing out key findings emerging from the comparison between the social protection models conceptualized and implemented 'from above' by public authorities with the models implemented 'from below' by workers own collective associations. This is followed by reflections on the power resources available to informal workers associations and the applicability of using the PRA to analyse these. Finally, the chapter discusses how the findings in the book point to a need for expanding the social protection agenda in several ways including suggestions of areas for further research.

Notes

1 Daladala are privately owned minibuses providing public transport.
2 Funded by the Danish Ministry of Foreign Affairs.
3 The basic needs development strategy as defined in the ILO report for the 1976 World Employment Conference to address 'basic needs' primarily through employment creation, targeted investments in health, education, shelter, access to water and sanitation, and the promotion of small-scale agricultural production grew out of the work of the ILO World Employment Program (WEP).
4 Methodologically, our book also differs from Awortwi and Walter-Drop (2018). For instance, in the latter, the case selection criteria are based on locations (districts) where the demand for social protection is particularly high and statehood is weak, thus implying some bias towards higher need areas (which might reinforce their rather unconditional celebratory view of informal social protection), whilst we select districts on the basis of different, more objective criteria (e.g. transport hubs) rather than social protection demand and thus we would expect more nuanced findings.
5 The PRA has its origins in the late 1960s and early 1970s where class became prominent again as an analytical category and also as a mobilizing principle among activists and left leaning scholars (Schmalz, Ludwig & Webster, 2018).
6 For examples of application of power resources to informal economy workers, see Kumar and Singh (2018); Riisgaard and Okinda (2018); Selwyn (2009); and Rizzo (2013).
7 Bodabodas are bicycle and motorcycle taxis commonly found in East Africa. In Kenya, they are more frequently called piki pikis.
8 In relation to contracts, we asked potential interviewees to specify first whether they had a contract and if so whether it specifies pay and entitlement to benefits, and if so whether the details of the contract are implemented in practice. If the answer to the first two were yes, but the answer to the last one was no, we proceeded with the interview on the basis of this being an informal worker.
9 In Tanzania, the construction sector sampling was also partly based on snowballing techniques in order to identify large and medium construction/building sites (see Chapter 3, Appendix B for the sampling method).
10 Based on our knowledge of existing informal worker organizations and initial fieldwork, approximately 30 informal worker organizations were sampled across all three sectors in both Kenya and Tanzania, covering different organization types (e.g. unions, cooperatives, and trade-specific).
11 After a thorough cleaning process, we ended up with a final sample of 1,385 observations. For further detail on the sampling method and data cleaning procedure, refer Chapter 3, Appendix B.

References

Adésínà, J.O. (2011) Beyond the social protection paradigm: Social policy in Africa's development. *Canadian Journal of Development Studies.* 32 (4), 454–470.

Alfers, L. & Moussié, R. (2019) Social dialogue towards more inclusive social protection: Informal workers & the struggle for a new social contract. In: *6th Conference of the Regulating for Decent Work Network.* Geneva, International Labour Organization.

Awortwi, N. & Walter-Drop, G. (eds.) (2018) *Non-state social protection actors and services in Africa: Governance below the state.* New York, Routledge.

Barrientos, A. (2010) Social protection and poverty. Poverty Reduction and Policy Regimes Thematic Paper, Social Policy and Development Programme No. 42. United Nations Research Institute for Social Development (UNRISD).

Bernards, N. (2017) The global governance of informal economies: The International Labour Organization in East Africa. *Third World Quarterly*, 38 (8), 1831–1846.

Brooks, S.M. (2015) Social protection for the poorest: The adoption of antipoverty cash transfer programs in the Global South. *Politics & Society.* 43 (4), 551–582.

Carré, F., Horn, P. & Bonner, C. (2018) *Collective bargaining by informal workers in the Global South: Where and how it takes place.* WIEGO Working Paper No. 38. Durban, WIEGO.

Chen, M.A. (2012) *The informal economy: Definitions, theories and polices.* WIEGO Working Paper No. 1. Durban, WIEGO.

Cherrier, C. (2016) *The expansion of basic social protection in low-income countries: An analysis of foreign aid actors' role in the emergence of social transfers in Sub-Saharan Africa.* PhD thesis. Maastricht University.

Chikarmane, P. & Narayan, L. (2015) *Organising the unorganised: A case study of the Kagad Kach Patra Kashtakari Panchayat (trade union of waste-pickers).* WIEGO.

de Soto, H. (1989) *The other path: The invisible revolution in the Third World.* New York, Harper and Row.

de Soto, H. (2001) *The mystery of capital: Why capitalism triumphs in the West and fails everywhere else.* New York, Basic Books.

Deacon, B. (2007) *Global social policy and governance.* London, Sage.

Deacon, B. (2013) *Global social policy in the making.* The Policy Press.

Devereux, S. & Sabates-Wheeler, R. (2004) *Transformative social protection.* IDS Working Paper 232. Brighton, Institute of Development Studies.

Devereux, S. & Sabates-Wheeler, R. (2007) Editorial introduction: Debating social protection. *IDS Bulletin.* 38 (3), 1–7.

DFID (2005) *Social transfers and chronic poverty: Emerging evidence and the challenge ahead: A DFID Practice Paper.* London, Department for International Development (DfID).

Dörre, K., Holst, H. & Nachtwey, O. (2009) Organising – A strategic option for trade union renewal? *International Journal of Action Research.* 5(1), 33–67.

du Toit, A. & Neves, D. (2014) The government of poverty and the arts of survival: Mobile and recombinant strategies at the margins of the South African economy. *The Journal of Peasant Studies.* 41 (4), 833–853.

Egan, G. (2014) *'Actually-existing' neoliberalism in Nairobi, Kenya: Examining informal traders' negotiations over access to the entrepreneurial city.* Master thesis. Carleton University, Ottawa, Ontario. Available from: https://curve.carleton.ca/system/files/theses/31951.pdf [Accessed April 2017].

Gentilini, U. & Omamo, S.W. (2011) Social protection 2.0: Exploring issues, evidence and debates in a globalizing world. *Food Policy.* 36 (3), 329–340.

Guha-Khasnobis, B., Kanbur, R. & Ostrom, E. (2007) *Linking the formal and informal economy: Concepts and policies.* WIDER Studies in Development Economics. Oxford, Oxford University Press.

Hanlon, J., Barrientos, A. & Hulme, D. (2010) *Just give money to the poor: The development revolution from the Global South.* Sterling, VA, Kumarian Press.

Hickey, S. & King, S. (2016) Understanding social accountability: Politics, power and building new social contracts. *The Journal of Development Studies.* 52, 1225–1240.

Hickey, S. & Seekings, J. (2017) *The global politics of social protection.* WIDER Working Paper 2017/115. Helsinki, The United Nations University World Institute for Development Economics Research (UNU-WIDER). Available from: doi:10.35188/UNU-WIDER/2017/339-4.

Hickey, S., Lavers, T., Niño-Zarazúa, M. & Seekings, J. (eds.) (2020) *The politics of social protection in Eastern and Southern Africa.* WIDER Studies in Development Economics. Oxford, Oxford University Press.

Holzmann, R. & Jorgensen, S. (1999) *Social protection as social risk management: Conceptual underpinnings for the social protection sector strategy paper.* Social Protection Discussion Paper Series, The World Bank. Washington D.C., The World Bank.

ILO (2015) *ILO recommendation 204: Transition from the informal to the formal economy.* Geneva, International Labour Organization. International Labour Office. Available from: https://www.ilo.org/dyn/normlex/en/f?p=NORMLEXPUB:12100:0::NO::P12100_ILO_CODE:R204.

ILO (2018) *Women and men in the informal economy: A statistical picture.* 3rd ed. Geneva, International Labour Office.

Jacob, T. & Pedersen, R.H. (2018) Social protection in an electorally competitive environment (1): The politics of Productive Social Safety Nets (PSSN) in Tanzania: ESID Working Paper 109. Manchester, UK, The Effective States and Inclusive Development (ESID) Research Centre.

Kakwani, N. & Subbarao, K. (2005) *Ageing and poverty in Africa and the role of social pensions.* Social Protection Discussion Paper Series. Washington D.C., World Bank.

Kamete, A.Y. (2018) Pernicious assimilation: Reframing the integration of the urban informal economy in Southern Africa. *Urban Geography.* 39 (2), 167–189.

KNBS (2019) *Economic survey 2019.* Nairobi, Kenya, Kenya National Bureau of Statistics.

Kinyanjui, M. (2012) *Vyama: Institutions of hope: Ordinary people's market coordination & society organization alternatives.* Oakville, Ontario, Nsema Publishers.

Kramon, E. (2019) *The local politics of social protection: Programmatic versus non-programmatic distributive politics in Kenya's cash transfer programmes.* International Growth Centre.

Kumar, S. & Singh, A.K. (2018) Securing, leveraging and sustaining power for street vendors in India. *Global Labour Journal.* 9 (2), 135–149.

Lavers, T. & Hickey, S. (2020) *Alternative routes to the institutionalisation of social transfers in Sub-Saharan Africa: Political survival strategies and transnational policy coalitions.* ESID Working Paper No. 138. Manchester, UK, The University of Manchester.

Li, T.M. (2007) *The will to improve: Governmentality, development and the practice of politics.* Durham, NC, Duke University Press.

Lindell, I. (2010) Introduction: The changing politics of informality-collective organizing, alliances and scale of engagement. In: Lindell, I. (ed.) *Africa's informal workers: Collective agency, alliances and transnational organizing in urban Africa.* London, Zed Publications, pp. 1–33.

Lourenco-Lindell, I. (2002) *Walking the tight rope: Informal livelihoods and social networks in a West African city.* PhD thesis. University of Stockholm, Stockholm Studies in Human Geography 9. Stockholm, Almqvist & Wiksell International.

Mkandawire, T. (2004) *Social policy in a development context.* London, Palgrave Macmillan.

Mkandawire, T. (2007) Transformative social policy and innovation in developing countries. *European Journal of Development Research.* 19 (1), 13–29.

Mitullah, W.V. (2010) Informal workers in Kenya and transnational organizing: Networking and leveraging resources. In: Lindell, I. (ed.) *Africa's informal workers: Collective agency, alliances and transnational organizing in urban Africa*. London, Zed Publications, pp. 184–206.

Marslev, K. (2019) *The political economy of social upgrading. A class-relational analysis of social and economic trajectories of the garment industries of Cambodia and Vietnam*. PhD thesis. Roskilde University.

Medina, L. & Schneider, F. (2019) *Shedding light on the shadow economy: A global database and the interaction with the official one*. The international platform of Ludwigs-Maximilians University's Center for Economic Studies and the ifo Institute (CESifo) Working Papers. Munich, Germany.

Nattrass, N.J. (1987) Street trading in Transkei – A struggle against poverty, persecution, and prosecution. *World Development*. 15 (7), 861–875.

NBS (2014) *Integrated labour force survey*. Dar es Salaam, National Bureau of Statistics.

Niño-Zarazúa, M., Barrientos, A., Hickey, S. & Hulme, D. (2012) Social protection in Sub-Saharan Africa. *World Development*. 40 (1), 163–176.

Oduro, A.D. (2010) Formal and informal social protection in sub-Saharan Africa. Paper Prepared for the ERD, August 2010.

Ouma, M., & Adésínà, J. (2018) Solutions, exclusion and influence: Exploring power relations in the adoption of social protection policies in Kenya. *Critical Social Policy*. Available from: doi:10.1177/0261018318817482.

Pal, K., Behrendt, C., Leger, L., Cichon, M. & Hagemejer, K. (2005) *Can low income countries afford basic social protection? First results of a modelling exercise*. Issues in Social Protection, No. 13, 2005, Social Security Department, International Labour Office. Available from: https://ssrn.com/abstract=807366.

Riisgaard, L. & Okinda, O. (2018) Changing labour power on smallholder tea farms in Kenya. *Competition and Change*. 22 (1), 41–62.

Rizzo, M. (2013) Informalisation and the end of trade unionism as we knew it? Dissenting remarks from a Tanzanian case study. *Review of African Political Economy*. 40 (136), 290–308.

Rizzo, M. (2017) *Taken for a ride: grounding neoliberalism, precarious labour, and public transport in an African metropolis*. Oxford, Oxford University Press.

Seekings, J. (2012) Pathways to redistribution: The emerging politics of social assistance across the Global 'South.' *Journal für Entwicklungspolitik*. XXXVIII (I–2012), 14–34.

Schmalz, S., Ludwig, C. & Webster, E. (2018) The power resources approach: Developments and challenges. *Global Labour Journal*. 9 (2), 113–134.

Selwyn, B. (2009) Labour flexibility in export horticulture: A case study of Northeast Brazilian grape production. *The Journal of Peasant Studies*. 36, 761–782.

Selwyn, B. (2014) *The global development crisis*. Polity, Cambridge.

Silver, B.J. (2003) *Forces of labor: Workers' movements and globalization since 1870*. Cambridge, Cambridge University Press.

Simpson, J.P. (2017) Do donors matter most? An analysis of conditional cash transfer adoption in Sub-Saharan Africa. *Global Social Policy*. Available from: doi:10.1177/1468018117741447.

Skinner, C. (2008) Street trade in Africa: A review. WIEGO Working Paper No. 5.

Steiler, I. (2018) What's in a word? The conceptual politics of 'informal' street trade in Dar es Salaam. *articulo —Journal of Urban Research*. 17–18.

Van de Walle, N. (2014) The democratization of clientelism in sub-Saharan Africa. In: Brun, D.A. & Diamond, L. (eds.) *Clientelism, social policy, and the quality of democracy.* Baltimore, MD, Johns Hopkins University Press, pp. 230–252.

von Gliszczynski, M. & Leisering, L. (2016) Constructing new global models of social security: How international organizations defined the field of social cash transfers in the 2000s. *Journal of Social Policy.* 45 (2), 325–343.

World Bank (2016) *Informal enterprises in Kenya.* Washington D.C., World Bank.

Wright, E.O. (2000) Working-class power, capitalist-class interests, and class compromise. *American Journal of Sociology.* 105 (4), 957–1002.

2 Formal social protection and informal workers in Kenya and Tanzania

From residual towards universal models?

Nina Torm, Godbertha Kinyondo, Winnie Mitullah and Lone Riisgaard

Introduction

In Kenya, following independence in 1963, the government has with its national development plans supported social protection as a key response to poverty through targeting the most vulnerable and desperate in society. However, the actual provision of social protection has taken many different forms and has generally been characterized by fragmented interventions, due in part to the various international agencies involved. In Tanzania, which gained independence in 1961 and has experienced a different economic history compared with Kenya, social protection has also consisted of fragmented efforts by a variety of actors, including international donors. As such, there is a lack of an overarching social protection policy framework that establishes clear institutional roles and responsibilities and lays out a concrete implementation plan from the central to local government levels. This compounded with a lack of financial resources and capacity both within government and civil society has until recently resulted in an inadequate provision of social protection for the general population.

However, in both countries, we are seeing a move towards more unified social protection approaches including the extension of coverage to informal workers. As already outlined in Chapter 1, this is partly a result of international recommendations, including the ILO's National Social Protection Floors (2012) and frameworks like the Sustainable Development Goals (SDG) putting pressure on African governments to expand the provision of social protection. At the regional level, social protection policy in both Kenya and Tanzania is governed by the African Union Social Policy framework which was endorsed by all Heads of States in 2009, and the wide-ranging provisions on the harmonization and coordination of social security that guide the actions of the East African Community (EAC) Partner States. For the purpose of this chapter, we will not go into detail on the global and regional governance of social protection yet concentrate on the relevant national frameworks and their implication for informal workers in Kenya and Tanzania, respectively. The rest of the chapter is composed of a section on Kenya's social protection framework with sub-sections focusing on the

DOI: 10.4324/9781003173694-2

National Social Security Fund (NSSF) and National Health Insurance Fund (NHIF), followed by a similar section on Tanzania zooming in on NSSF and NHIF/Community Health Fund (CHF). A brief sum-up is provided at the end of the chapter.

Kenya's social protection framework

In 2006, under the auspices of the African Union, the government of Kenya began the process of formulating a national social protection framework in consultation with representatives from government ministries, non-state actors including the private sector, community groups, voluntary organizations, and development partners, whilst also exploring international best practices in the provision and financing of social protection. The resulting National Social Protection Policy (NSPP, 2011) outlines how the consultations identified several key barriers that were preventing people from accessing social protection services, including stigma and discrimination on account of gender, disability, age, nationality, area of residence, and poor wellbeing.[1] Yet, among these barriers to accessing social protection services, informality is not explicitly spelled out.

The NSPP defines social protection as follows:

> Policies and actions, including legislative measures that enhance the capability of and opportunities for the poor and vulnerable to improve and sustain their lives, livelihoods, and welfare, that enable income-earners and their dependants to maintain a reasonable level of income through decent work, and that ensures access to affordable healthcare, social security, and social assistance.
>
> (NSPP, 2011, p. v)

The NSPP is consistent with Kenya's 2010 constitution which in article 43 expressly guarantees all Kenyans their economic, social, and cultural (ESC) rights including basic rights to health, education, food, and decent livelihoods. It explicitly asserts the right "of every person... to social security" and binds the State in Article 43(3) to "provide appropriate social security to persons who are unable to support themselves and their dependants".[2] The right to social security, in both the wide and narrow sense, is closely interlinked with other social protection rights including the right to healthcare services, equality and freedom from discrimination, human dignity, freedom of movement and residence, reasonable working conditions, fair administrative actions, access to justice, and the resolution of disputes in a fair manner and through public hearing before a court or independent and impartial tribunal or body (NSPP, 2011, p. 1). As such, social protection is geared towards improving human capabilities through social assistance (cash transfer programmes) and/or social security (pensions and health insurance). In terms of the latter and in relation to informality, the NSPP states that

The Government, working with all other stakeholders, shall:

i Strengthen the existing social security regime and establish comprehensive social security arrangements that will extend legal coverage to all workers, whether in the formal or informal sectors, and their dependants (NSPP, 2011, p. vi).

ii Undertake research into and consider viable options for extending coverage to those who work informally and their dependants in consultation with key stakeholders, including those in affected communities and sectors (NSPP, 2011, p. vii).

Interestingly, in the first paragraph, the wording *informal sector* gives the impression of a limited segment of informal workers, whilst the second paragraph refers to *those who work informally* which is a broader category of workers in the informal economy. Regardless of the subject, the cited objectives on extending social protection measures to the informal are clearly in line with the ILO agenda of "formalizing the informal" (ILO, 2015).

The NSPP is also in line with Kenya's Vision 2030 which aims to provide a "high quality of life for all its citizens by the year 2030". Built on three pillars – economic, social, and political, the social pillar of the Vision seeks to build "a just and cohesive society with social equity in a clean and secure environment". Vision 2030 is the national long-term development policy which has provided an impetus for the government to increase investment in social protection to the levels invested in comparable countries. For the first five-year period (2008–2012), the Vision's goal was "to increase opportunities all-round among women, youth, and all disadvantaged groups", and one of the actions proposed to achieve this was the establishment of a consolidated Social Protection Fund administered by the National Social Protection Council (NSPC). Whilst there has been limited progress in reforming the contributory schemes (KSPSR, 2017), progress has been made along other parameters, including the senior citizens "Inua Jamii 70 years and above" programme which was introduced in January 2018, representing Kenya's first universal social protection programme, guaranteeing a minimum pension to all older persons aged above 70 years. The scheme is funded by the government and has been rolled out to around 533,000 senior citizens, expanding the Older Persons Cash Transfer programme (OPCT), which was targeted to poor and vulnerable households. If "Inua Jamii 70 years and above" attains full coverage, around 840,000 persons will be able to access the scheme and about 68,000 people aged 65–69 years will continue on the OPCT. In total, it is expected that the share of those aged 65 years and over who receive an old age pension should increase to 77 percent from around 31 percent (KSPSR, 2017). Aside from being a response to the limitations of targeting, this more universalistic approach is portrayed as an effective solution to offering pensions for the informal economy.

The programme is expected to address many of the current challenges with pension coverage some of which are outlined later on in the NSSF section.

With regard to the health insurance pillar of social protection, the NSPP does not explicitly mention informal workers but states that the government, in collaboration with partners, shall:

- i Re-establish the NHIF as a fully-fledged comprehensive national health insurance scheme, which covers all Kenyans, and to which those who can afford it must contribute (NSPP, 2011, p. vii).
- ii Establish a framework for enabling those who are not able to contribute to access a core package of essential health services, including maternity care and treatment for HIV/AIDS and related diseases (NSPP, 2011, p. vii).

Following the NSPP, there have been two Social Protection Sector Reviews (KSPSR, 2012, 2017), and during this period, both the NSSF and the NHIF have opened up to informal workers. The most recent KSPSR (2017) lists the progress made in this regard, and the key dimensions are summarized in Table 2.1.[3]

From the brief overview, it is thus clear that there has been limited progress in terms of extending both pensions and health insurance to informal workers, and as indicated earlier, the underlying discourse remains one of *formalizing the informal.* More specifically, when relating the recommendations from the KSPSR (2012) to the different trajectories on social protection, as outlined in Chapter 1,

Table 2.1 Summary of progress in implementing key recommendations from the KSPSR, 2012, related to the informal economy

Proposal	*Progress*
To strengthen the existing social security regime and establish comprehensive social security arrangements that will extend legal coverage to all workers, whether in the formal or informal sectors, and their dependants.	Limited progress. However, the universal *Inua Jamii Senior Citizens'* programme is considered as a more effective tool for offering workers in both formal and informal sectors a pension. It will need to be established in law for it to be considered as a legal coverage.
Determine the most appropriate role to be played by occupational schemes in extending social security coverage to those who can contribute to their own post-retirement welfare and security and risk mitigation.	There has been some expansion of contributory retirement schemes for those in the informal economy, but they still provide only lump sums so act more like savings schemes (see Box 2.1 on the Mbao Pension Plan).
A national social health insurance scheme will be initiated that will protect both formal and informal sector workers as well as the unemployed from the economic liability of health shocks.	Limited progress. The focus is still on the NHIF, and a broader health insurance scheme has not been formally introduced. However, through the Universal Health Coverage (UHC) programme in four counties (Kisumu, Nyeri, Isiolo, and Machakos) the government is working on expanding social protection to informal workers.

Source: Adapted from KSPSR (2017).

they can be viewed as being located somewhere between the left (beginning) of the continuum towards the middle of the continuum. Whilst the proposal on contributory retirement schemes for risk mitigation and health schemes to address the economic liability of shocks are more in line with the residual approach of, for instance, the World Bank, the recommendation to "extend legal coverage to all workers, whether in the formal or informal sectors, and their dependants" resonates more with the perception of social protection as a right in accordance with universal models. In that sense, there is still very limited recognition of the broader transformative role of social protection as emphasized in the Devereux and Sabates-Wheeler (2004) framework, not to mention the plurality of informal social protection measures as revealed throughout this book. This disconnect between the official social protection policy stance, its implementation and the reality on the ground is where our research is situated. In what follows, and with the purpose of providing the context for the subsequent chapters, we provide an account of the two main public insurance schemes in Kenya and how they relate to informal workers.

National Social Security Fund

When the NSSF was established through an Act of Parliament in 1965, it only covered men that were formally employed; yet, in 1977, the scheme opened up to women and with the 2013 NSSF Act coverage was extended to workers in the informal economy. Today, the NSSF covers all categories of employers and workers and is a mandatory retirement scheme whose main objective is to provide basic financial security benefits to Kenyans in both the formal and informal sectors (NSPP, 2011, p. 12). The NSSF has two tiers of contribution, namely the provident fund and the pension fund, and whilst the latter is obligatory for all formal employees from the age of 18–60, the former is voluntary and caters for self-employed and unemployed workers who fail to meet the minimum provisions. Prior to 2013 contributions amounted to 5 percent of earnings by each of the employee and employer, subject to a cap on earnings of Kenyan Shillings (KES) 400 (USD 4) per month. Currently, contributions, which depend on minimum wages, amount to 6 percent payable by each of the employee and employer with the 'cap' having increased to KES 2,160 per month (USD 21) (KSPSR, 2017). In return, members receive benefits upon turning 60 years. Although NSSF annual contributions and annual benefits increased by 5.2 percent and 2.6 percent, respectively, in 2018 (KNBS, 2019), the low monetary ceiling on contributions means that the current level of benefits is generally considered inadequate.

In terms of coverage, during 2014–2018, the number of registered employers and employees increased by 6.8 percent and 3.0 percent to 143,300 and 4,068,400, respectively. During the same period, the number of registered women employees increased by 4.8 percent compared to 2.4 percent growth recorded for men employees (KNBS, 2019). Despite such increases and the fact that the NSSF has 60 branches and is accessible in each of the 47 counties through 'Huduma Centres', effective coverage remains low. In terms of active members, it is estimated that, for formal *and* informal workers, only 15 percent of those aged 18–65 years have an employer contributing to NSSF pensions (KSPSR, 2017), which is nevertheless an increase from 2011 (NSPP, 2011, p. 12). The low level of benefits could be a factor dissuading especially informal workers from joining the scheme. Moreover, as discussed throughout this book, low enrolment of informal workers is attributed to challenges such as low wages, lack of contracts, high poverty, little awareness of NSSF, and the bureaucratic nature of the scheme. In addition, scant information on informal economy workers makes it difficult for government programmes to prioritize them; yet, in order to extend coverage further, the NSSF is working with informal worker organizations, examples of which are provided in the subsequent chapters of this edited volume.

Since 2013, NSSF members have the option of 'contracting out' of the mandated pension component if the individual worker has a pension through an alternative scheme. An example is the 'Mbao' scheme (Box 2.1) which is an initiative to extend private pensions to the informal sector; however, in line with existing literature (Kabare, 2018), our research shows that very few informal workers subscribe to Mbao. Moreover, it appears that the scheme is not well understood. For instance, the phrase "'tupa' (throw) twenty shillings" which was used by some workers when talking about the scheme revealed the suspicion of those throwing the 20 shillings.[4] Observations also revealed that the Mbao

Box 2.1 The 'Mbao' scheme: a flagship initiative of the Retirement Benefits Authority (RBA)

The Mbao Pension Plan was established in 2009 by the Kenya National Federation of Jua Kali Associations as a voluntary retirement savings scheme, geared towards supporting citizens engaged in the informal sector who are not accessing any social security support, although it is open to any citizen who would like to join. It is registered as a retirement benefit scheme, yet is managed privately in the same way as most of the 'normal' occupational schemes operating in the country. Mbao offers members to easily accumulate a low level of regular savings at minimal cost – typically though the money transfer schemes operated by the mobile phone companies. In 2016, the total recorded membership was about 99,000, with an accumulated fund of about KES 110 million. Thus, while the scheme membership has, until now, reached only a rather limited proportion of the target population of informal economy workers, it has grown fairly steadily. However, since contributions can be – and typically appear to be – withdrawn after a required minimum period of three years, there is little evidence that Mbao will be able to contribute to old age income security. Thus, whilst being modestly attractive as a savings vehicle, it is unlikely to function as a reliable pension scheme for those working in the informal economy (KSPSR, 2017).

scheme is 'a child without a parent', with many agencies claiming ownership including the Jua Kali Association, Kenya Commercial Bank, Kenya Retirement Benefits Authority, Eco Pension Plan, and the ILO. Thus, governance of the Mbao scheme seems not well consolidated. In January 2020, the NSSF in collaboration with the NHIF, Kenya Commercial Bank, Safaricom, and Proto Energy launched a new pension scheme for informal workers named 'Haba Haba', seeking to address the immediate and medium term needs of present and future NSSF members, initially targeting informal worker in the transport sector. The scheme gives members of the informal economy a chance to save a minimum of KES 25 (USD 0.25) a day with the option of withdrawing 50 percent of their contributions after consistently contributing for a minimum of five years. This differs from NSSF offers in the formal sector where members have to retire or attain the age of 50 years to access their savings. At the time of writing, it is, however, unclear how or whether the 'Haba Haba' scheme relates to the Mbao scheme and whether the latter still remains in operation.

National Hospital Insurance Fund

The NHIF was established in 1966 as a department under the Ministry of Health, and like NSSF, it was part of the government's efforts to cushion workers against future vulnerabilities. Although several schemes in Kenya offer healthcare, the NHIF is the primary one covering more than 90 percent of the insured population. In 1972, the NHIF extended coverage to self-employed/informal workers on a voluntary basis, and later on, outpatient coverage for both formal and informal sector workers was introduced in addition to in-patient services (NSPP, 2011, p. 13). In 2012, the NHIF established a health insurance package for informal workers providing beneficiaries with access to a range of treatments and services for a monthly premium contribution of KES 500 (USD 4.6) per family paid by the individual worker. Moreover, everyone who receives the "Inua Jamii 70 years and above" pension is also accepted as members of the NHIF, with their contributions paid by the government (KNBS, 2019).[5]

In terms of the number of NHIF registered members, this has increased gradually from 4.5 million in 2013/2014 to 7.7 million in 2017/2018 (KNBS, 2019) and 9 million in 2019/2020 (KII, Nairobi).[6] Since the NHIF also provides for immediate family members, the total number of people benefiting was around 18.4 million in 2017, equivalent to around 39 percent of the population (KSPSR, 2017).[7] The rate of membership expansion has been particularly high among informal workers who saw a 23 percent rise for the period 2013/2014 to 2017/2018 compared with a formal sector increase of 4 percent (KNBS, 2019). In terms of numbers, informal workers made up around 700,000 contributing members in 2011 increasing to an estimated 2.5 million in 2015/2016 (KSPSR, 2017) and 3.2 million in 2019/2020 (KII, Nairobi). Thus, around 41 percent of total members come from the informal economy; yet, considering that there are around 14 million informal workers, the coverage gap among this group remains substantial.

As indicated in Table 2.1, the UHC programme, which was launched at the end of 2018, presents an attempt to further expand social protection to informal workers.[8] The programme was supposed to operate for one year after which a review would assess how to roll it out to the rest of the 46 counties; yet, it was extended for three months with no extra resources. During this time, 40 other counties had signed the Intergovernmental Participation Agreement (IPA) with the national government. Further commitment was made through an inter-governmental relations agreement in October 2020, with provisions for developing a UHC policy and review of the NHIF to provide for UHC, including establishment of a UHC Fund (ROK, 2020). All these commitments are in progress as the government intensifies provision of primary health care, viewed as a foundation for UHC. However, public observations indicate that the programme has been underfunded with some pilot counties not even receiving half of the budgeted amount. Moreover, even if/when the programme is rolled out country-wide, it seems likely that, given the low sustainability and predictability of informal workers' income, the majority of premium contributors will not be consistent in their payment and will likely require subsidies (Okungu & McIntyre, 2019). In fact, the inconsistency of informal sector contributions has prompted the NHIF to put in place a two-month restriction on usage of the NHIF card for self-employed contributors to avoid people only enrolling when they are seeking service.

The research presented in this book reveals that some of the reasons for poor enrolment/inactiveness of informal workers include cost of premium, registration problems, lack of information, and informal associational cushioning. In order to better reach informal economy workers, the NHIF has started to engage them through their associations, something which will be discussed in the sector-specific chapters. Although, the number of informal economy workers that have joined the NHIF remains relatively low, the NHIF has, until now, been more successful than the NSSF.

Tanzania's social protection framework

Tanzania's Development Vision 2025, the National Social Security Policy NSPP (2003), the National Ageing Policy (2003), and the Health Sector Strategic Plan 2015–2020 all envisage social protection for every citizen in Tanzania. Towards this purpose and following the publication of the NSPP in 2003, the Tanzanian government has enacted a number of legislations to address social security issues, leading to the establishment of the Social Security Regulatory Authority (SSRA) in 2008, with the mandate to govern and regulate the pensions and health insurance schemes. In 2010, from an initial draft in 2000, Tanzania published its National Social Protection Framework (NSPF).

> NSPF (the National Social Protection Framework) is part of national efforts to reduce poverty and its primary aim is to reach the most vulnerable and ensure their protection. It is also a means of building the capabilities of the

poor to engage in production so that they become effective participants in and beneficiaries of the growth process.

(United Republic of Tanzania, 2008, p. 2)

The wording in this statement is very much along the lines of the residual risk management approach to social protection, where the latter is not seen to play a developmental role in itself, but rather, there is a general trust in the ability of the market to deliver broad economic ***and*** social progress. Following the NSPF, in 2012, the social security law (SSL) was amended, and coverage extended to "apply to any person employed in the formal or informal sector or self-employed within mainland Tanzania" (SSL, 2012, p. 50). The amended SSL also defined the informal sector as "the sector which includes workers who work informally and who do not work in terms of an employment contract or any other contract contemplated in the definition of employee" (SSL, 2012, p. 51). Initially, the 2012 SSL banned the withdrawal of benefits before retirement age; yet, after public outcry, the parliament lifted the ban later that year. However, with the passing of the Social Security Fund Act in 2018, the ban is back so as to preserve the meaning of pensions, and a member who resigns will have to wait until they reach the age of 55–60 to get their benefits.[9]

Despite their wording, neither the 2003 policy nor the 2010 social protection framework sufficiently addressed issues of informal workers and participation of poor households, and neither define clearly how different actors could implement social assistance programs and voluntary schemes. These shortfalls in design, implementation, coordination, monitoring, and evaluation processes for linking up and ensuring national coverage in the implementation are still under discussion. Since April 2018, a new NSPP has been proposed consisting of three pillars: social assistance, social insurance (health and pension), and voluntary schemes.[10] The latter would cover associations of informal workers for instance. Whilst the revised NSPP is still to be approved, three public insurance schemes – the NSSF, the NHIF, and the CHF – are all open to informal workers and in what follows we consider the relevance and potential accessibility of each of these.

National Social Security Fund

The NSSF caters for employees in the private sector, the self-employed, foreigners employed in Mainland Tanzania, and employees in international organizations based in Mainland Tanzania. Any category of temporary employee is registrable under the NSSF, and the "The National Informal Sector Scheme" applies specifically to informal workers. The informal membership figures have varied substantially year-on-year and, in 2017/2018, registered informal worker members represented 18,631 down from 47,780 in 2014/2015 (NSSF, 2018).[11] One explanation for the sharp fall in informal members is that whilst workers are initially attracted to joining the scheme in order to receive loans and other short-term benefits, in the longer run, they drop out and/or join informal groups for the reasons discussed in the sector specific chapters. In addition to social security, the NSSF offers health and medical coverage, however this benefit is seemingly on hold due to the high administrative expenditure at around 22 percent of the Fund's income. In early 2018, the Public Service Social Security fund (PSSSF) was signed into Law including a Voluntary Savings Retirement Scheme (VSRS) for the employed, self-employed, informal sector workers, and all other categories of people such as farmers, fishermen, drivers, micro-traders, food venders, organized groups such as VIBINDO (an association of small businesses), Saving Associations and Credit Co-operative Societies (SACCOS), women groups, and professionals.[12] Under the VSRS, contributors were eligible for pension and health insurance upon payment of TZS 20,000 per month during three consecutive months; however, the health insurance part was deemed too costly and was not renewed after the end of 2019.

Despite these openings, informal sector workers are largely excluded from the ambit of the social security legislative framework, due to a number of challenges outlined by Ackson & Masabo (2013). First, all the schemes (except CHF) require contributions from both the employer and the employee, and workers in the informal sector may be unable to contribute "double" portions on behalf of their "non-existent" employer and themselves. Even when informal workers do have employers, in the absence of a regulatory framework, employers may be unwilling to make contributions to a social security fund because this increases their labour costs. Similarly, in cases where employer contributions are not mandated by law, informal workers may be reluctant to allow their employer to deduct contributions to the social security scheme as they would receive lower pay. Moreover, the benefits of the schemes are not deemed to be substantial enough to warrant contributions, whilst uncertainty and unreliability of income may impair informal sector membership, contributions, and qualifications for benefits. For instance, for a member to qualify for invalidity pension and maternity benefits under the NSSF, workers must have contributed for a minimum of 36 months out of which 12 months contributions must have been made within the immediate past 36 months before the occurrence of the social risk.[13] Finally, workers in the informal sector are unaware of the core functions and operations of social security schemes. In order to address some of these issues, the NSSF has been providing short-term benefits to informal sector contributors as a motivating factor for them to continue their membership. For example, each year, upon satisfactory contribution from an informal sector member, the NSSF will issue that member with an appropriate asset – i.e. fuel or tyres for bodaboda workers, or cooking utensils, chairs, etc., for informal food vendors.

National Health Insurance Fund and Community Health Fund

The largest health insurance schemes are the NHIF and the CHF/iCHF. From the early 1990s, the Tanzanian government adopted reforms that changed the health financing system from free services to mixed financing; yet, health insurances were only formally introduced in 2000, as part of larger health reforms aimed at improving access to health services. By then, it was becoming increasingly difficult for ruling politicians to defend the inconsistency between the promise of universal access to social services heralded under African socialism and the reality of limited access (Pedersen & Jacob, 2018).[14] Whilst the NHIF was initially meant for civil servants, employees in executive agencies and government parastatals, coverage was later on extended to private-sector employees and the self-employed plus family members. Following the health-sector reforms in the early 1990s, the CHF was introduced in 1996 as part of the endeavours to make health care affordable and available to the rural population and workers in the informal sector on a voluntary basis. In 2001, the government passed the CHF Act (supported by the World Bank), formally institutionalizing it as a voluntary insurance-based hybrid scheme, administered at the district level and co-financed by the community (household) and the government. In practice, this means that the CHF members must first seek care at a dispensary or health centre in their districts which then submits the claims to the central government that then matches the collected membership fees. In 2009, the NHIF was contracted to oversee operations of the CHF which targets the rural informal population and its counterpart TIKA which covers the informal sector in urban areas (Mbekeani, 2009; Rwegoshora, 2016; Pedersen & Jacob, 2018).

The CHF has encountered challenges including poor enrolment and poor health services offered to members. Since 2014, the CHF has been under replacement by an improved CHF system (iCHF), permitting members to access health services outside of their districts and providing a more comprehensive package of services. Thus, the iCHF was introduced as a voluntary, district-owned health insurance scheme, built on a strong partnership between the NHIF, the district councils, public and private health care facilities, and PharmAccess. By September 2016, more than 100,000 people had enrolled in the iCHF, and the aim is to enrol at least 30 percent of the population in each of the districts of the Northern Zone's Kilimanjaro, Arusha, and Manyara regions by 2021. The government has also successfully tested the iCHF in the regions of Dodoma, Kilimanjaro, and Mbeya. In 2019, after assuming oversight of the CHF/TIKA schemes, the Ministry of Health, Community Development, Gender, Elderly and Children (MOHCDGEC) assessed lessons learned and has taken initial steps to consolidate and rename the schemes referring now to all former CHF/TIKA schemes collectively as the iCHF scheme. THE MOHCDGEC has also begun to pool funds at the regional level to enable greater cross subsidization of health costs. The new standard benefit package for the iCHF centres on the provisions of primary care; yet, benefits are more limited than those offered by the NHIF and most private health insurance schemes. The iCHF charges a single per-beneficiary premium of TZS 40,000 (USD 20) annually in Dar es Salaam or TZS 150,000 (USD 65) per household of six and TZS 30,000 (USD 13) in all other urban areas. In rural areas, premiums range from TZS 10,000–30,000 with the latter being for a household of six, and for all locations, the government matches the contribution (Lee, Tarimo & Dutta, 2018). The parliament is expected to pass legislation in late 2020/early 2021 that will authorize the national rollout of iCHF as preparation for the Universal Health Care policy.

As for the NHIF, individual households from the informal sector can voluntarily enrol for an annual premium of approximately USD 672, which is well above the iCHF and beyond the means of many informal workers.[15] In terms of membership, the NHIF recorded an increase in their membership base from 164,708 in 2001/2002 to 468,611 in 2010/2011 up to around 750,000 in 2017 (Prabhakaran & Dutta, 2017). Since up to six family members can be added to the employee's membership, this means that the total number of beneficiaries might be about 3.5 million, equivalent to around 7 percent of the population (Wang et al., 2018).

This is in line with estimates provided by Pedersen and Jacob (2018) according to whom existing insurance schemes covered approximately 22 percent of the population (7 percent by the NHIF, 12 percent covered by the voluntary CHF, 1 percent by private insurances, 1 percent by community-based health insurances and 1 percent by the NSSF).[16] According to NHIFs own data, coverage has been steady around 7 percent of the population in the period from 2012/2013 to 2017/2018, whereas a large increase is reported for CHF for the same period (Riisgaard, 2020). Thus, as of 2018, it is estimated that around 34 percent of Tanzanians have some form of health insurance with 25 percent (13.5 million beneficiaries) covered by the CHF, 7 percent by the NHIF with private health insurance accounting for 2 percent (NHIF Factsheet for FY2017/18).[17] This rise in CHF enrolment is also reported by other sources (Wang et al., 2018) citing an increase in the membership base from around 7.4 percent of the population in 2011 to 19.8 percent in 2015. In actual numbers, this means that, as of June 2017, more than 2 million households were enrolled in the CHF serving over 12 million beneficiaries.

In 2015/2016, several new schemes were introduced under the NHIF attempting to broaden the uptake. This included a scheme for the elderly ('Wazee Kwanza') for children under 18 years ('Toto Afya Kadi') and for expectant mothers ('Mama na Mwana') and the NHIF mutual plan, known as the KIKOA scheme which is a health insurance scheme for groups in the informal sector e.g. VIBINDO, SACCOS, AMCOS, VIKOBA and special groups of registered entrepreneurs like motor cycle drivers (boda boda), food vendors or any other entrepreneurial group of members not less than 20.

The KIKOA scheme extended informal sector access to the NHIF through allowing members of registered associations in the informal economy with a minimum of 20 members (who had been together for a minimum of six months) to sign up for TZS 76,800 (USD 35) annually per person. In addition, members can add up to six dependents for an additional annual fee of TZS 76,800 per head, thus substantially above the CHF cost. According to 2017/2018 figures from the NHIF, the KIKOA scheme reached 33,057 members, equivalent to 3.85 percent of the total NHIF membership. Although, in theory, such a group-based insurance NHIF scheme seemed well-suited and affordable, it appears, from the research presented in this book, that the membership cost has been a significant burden for many. In addition, for the more insecure and weaker groups, the NHIF does not even appear to be on their radar. These findings are echoed in the literature (Mushi & Millanzi, 2019), and many of the issues are similar to those of informal workers in Kenya, as seen earlier. For reasons related to adverse selection issues, the KIKOA program was terminated by NHIF towards the end of 2019, and instead, the NHIF has other models such as the bodaboda Afya (health) which costs TZS 100,000 per head.

Private institutions such as the Jubilee Insurance have previously had health insurance schemes targeting informal workers, and other smaller private-sector health insurance providers also exist, along with various informal micro-health insurance schemes provided by, for instance, churches and cooperatives mostly operating on a very small scale (Mills et al., 2012; Pedersen & Jacob, 2018). To this should be added the myriad of different formal and informal associations which, as will be shown in later chapters, offer a one-off amount in case of health-related challenges. Given that only 3 percent of total health financing is covered by official insurance schemes (Pedersen & Jacob, 2018) with the remainder being out of pocket, the importance of such informal health insurance schemes should not be underestimated. This is particularly relevant for poorer sections of the population as, according to Wang et al. (2018), demographic and health survey data from the period 2015–2016 shows that NHIF covers mostly the top 40 percent of the income distribution.

As in other SSA countries, the influence of international development partners in the adoption of health insurances is undeniable and largely mirrors the international trends described earlier. However, as shown by Pedersen & Jacob (2018), major reforms (in both social assistance and in healthcare) were typically introduced with some very Tanzanian characteristics. For health insurances, a characteristic feature

> was the way in which a policy coalition of bureaucrats and development partners framed reforms as a way for the ruling party to live up to one of its core priorities since independence, namely improved and, eventually, universal access to health services, while at the same time improving the efficiency of the sector through the introduction of fees and insurances. The latter were also seen as a way to mobilise additional funds for the sector.
>
> (Pedersen & Jacob, 2018, p. 4)

In general, the idea of universal access to social services – a prominent feature of the *Ujamaa* ideology has re-emerged among ruling politicians in the 2000s, especially in the wake of the hardship brought about by the liberalizing economic reforms of the 1980s. This can be seen in the Vision 2025 development programme, which heralded the reintroduction of universal primary and secondary school education which has also spread to the health sector (Pedersen & Jacob, 2018, p. 12). The National Health Finance Strategy for the period 2015–2020 proposes a single national health insurance (SNHI) program to be managed by the NHIF, which will be tasked with strengthening the iCHF/TIKA in order to reach low-income people and the informal sector. The Parliament was expected to pass SNHI legislation in 2017 (Wang et al., 2018); however, the law has not yet been approved. The interim plan while waiting for the SNHI is to have two coexisting schemes: the NHIF aimed at covering the formal sector and the iCHF intended to cover the informal sector and rural households (Lee, Tarimo & Dutta, 2018). According to Pedersen and Jacob (2018), so far, however, the route taken by the ruling party has been focused more on improving public health infrastructure than health insurances. In addition, even though reforms in health insurance have been pursued, the single mandatory national health insurance heavily pushed by development agencies has so far not materialized.

In summing up, despite their different economic histories, it is clear that, when it comes to social protection, Kenya and Tanzania have similar experiences and trajectories, with both countries currently seeing a very gradual shift from the more residual end towards the more universal end of the social protection spectrum. In both countries, the limited informal worker uptake of formal social protection schemes is linked to issues such as the cost burden and lack of information on the different options. In Tanzania, the iCHF, which is a step towards the roll-out of Universal health insurance, is carried out at the local government level in order to acclimatize the population to the importance of having health insurance. Eventually, after merging the NHIF, iCHF, and NSSF, Tanzania will have a SNHI managed by the NHIF. Thus, although universal health coverage has become a major policy priority in Tanzania, progress on this key aspect of the SDGs is further ahead in Kenya where 2020 saw universal health insurance pilot programs being carried out in four counties. Similarly, Kenya's universal pension scheme Inua Jamii Senior Citizens launched in 2018 is fully government funded, whereas Tanzania's VSRS, also from 2018, as the name indicates relies on contributions from informal workers and their (often non-existent) employer, as such limiting its effectiveness and sustainability. In both countries, it remains to be seen whether the progress towards more universal models of social protection will be stalled or even reversed with the ongoing COVID-19 crisis.

Notes

1 At the time of writing (end of 2020), the Kenyan government was reviewing the NSPP.
2 Wanyama and McCord (2017) look at the role of Kenya's political settlement in the adoption and promotion of social protection.
3 The KSPSR (2017) calls for further analysis to determine how the contributory schemes can include a higher proportion of people working in the informal economy and whether this would require further changes in legislation.
4 The concept of tupa' is an abstraction of the word 'usitupe' which means "do not throw away".
5 As for salaried employees, contributions are compulsory, with premiums calculated on a graduated scale (between KES 150–1,700/USD 1–16) based on the employees' income and subtracted directly from the payroll.
6 However, it is estimated that only around 4 million NHIF members (out of the 6.2 million) are consistent with their payments (Waweru, 2017). Aside from inactive membership, the NHIF faces a number of other challenges, including reimbursement policies, which have encouraged longer stays in hospitals, the increase in the value of claims, an uneven distribution of payments in different categories of hospitals, and reimbursements that are skewed in favour of private hospitals and nursing homes

rather than government or mission facilities. Finally, there are also administrative inefficiencies that have contributed to high overhead costs.

7 Mwaura et al. (2015) estimate that private, microfinance and community-based health insurance cover 2 percent of the population.

8 Another strategy for reaching informal workers is the Health Insurance Subsidy Programme (HISP) which is enabling the members of the CT-OVC (Cash Transfer for Orphans and Vulnerable Children) and OPCT schemes to access the NHIF. A 'block' premium is paid by the government to the NHIF corresponding to that payable by the minimum-rate voluntary contributors. This arrangement has been pre-trialled for a limited number of households, with external donor support. However, no information is available on whether assessments have been made of the long-term financial sustainability of this arrangement.

9 If a member loses his/her job either by being fired or retrenched, then that person receives 33 percent of the equivalent of the last salary for a period of six months ('unemployment benefit') and waits for up to two years during a probation period after which it is possible to request withdrawing the benefits and be shifted from the mandatory to a voluntary-based scheme where the applicant pays his or her own benefits.

10 The proposed new policy has, at the time of writing (early 2021), not yet been submitted to cabinet for approval, and there is no official explanation from the government. Matters have been complicated by the movement of government staff from Dar es Salaam to the new capital Dodoma and the dissolving of the SSRA. Whilst official government sources (directorate of social protection in the Ministry of labour) informed that the Trade Union Congress of Tanzania (TUCTA) was involved in the formulation of the policy, TUCTA claims that their inputs were sought *after* the formulation of the document thus questioning whether their opinions really mattered at that juncture (interview with the former Secretary General of TUCTA).

11 The total registered number of NSSF members was 247,000 in 2020 yet counting only 11,000 active members.

12 Besides the NSSF, the public pensions schemes included the Parastatal Pensions Fund (PPF), the Public Service Pensions Fund (PSPF), the Government Employees' Provident Fund (GEPF), and the Local Authorities' Pensions Fund (LAPF); however, with the PSSSF Act in 2018, all of these were dissolved and combined into one general fund to cater for the public sector.

13 The criteria that one has to contribute for 180 consecutive months (15 years) to be eligible for pension under the current NSSF pension scheme makes it a very poor fit with the reality of most informal traders (Riisgaard, 2020).

14 See Pedersen and Jacob (2018) for a detailed account and analysis of the introduction and expansion of health insurance schemes in Tanzania.

15 As for NHIF contribution rates by formal sector workers, these are fixed at 6 percent of the total salary, split equally between the employer and employee.

16 Pedersen and Jacob (2018) also note that figures on health insurance coverage are uncertain as they vary for similar time-periods in different documents. For instance, Lee, Tarimo, and Dutta (2016) report 20 percent coverage for 2014/2015.

17 While 60 percent of health insurance schemes belong to private commercial companies, they cover a very small part of the population. Among the most prominent are AAR, Jubilee, Resolution, and Strategis.

References

Ackson, T. & Masabo, J. (2013) *Social protection for the informal sector in Tanzania*. Available from: http://www.saspen.org/conferences/informal2013/Paper_Ackson-Masabo_FES-SASPEN-16SEP2013-INT-CONF-SP4IE.pdf.

Devereux, S. & Sabates-Wheeler, R. (2004) *Transformative social protection.* IDS Working Paper 232. Brighton, Institute of Development Studies.

ILO (2012) *ILO recommendation 202: Social protection floors.* International Labour Organization. Geneva, International Labour Office. Available from: https://www.ilo.org/dyn/normlex/en/f?p=NORMLEXPUB:12100:0::NO::P12100_ILO_CODE:R202.

ILO (2015) *ILO recommendation 204: Transition from the informal to the formal economy.* International Labour Organization. Geneva, International Labour Office. Available from: https://www.ilo.org/dyn/normlex/en/f?p=NORMLEXPUB:12100:0::NO::P12100_ILO_CODE:R204.

Kabare, K. (2018) The Mbao pension plan: Savings for the informal-sector. *Development pathways.* Working Paper: October 2018. Nairobi, Kenya.

KNBS (2019) *Economic survey 2019.* Nairobi, Kenya, Kenya National Bureau of Statistics.

KSPSR (2012) *Kenya social protection sector review 2012.* Nairobi, Kenya, Ministry of Labour and Social Protection, State Department for Social Protection.

KSPSR (2017) *Kenya social protection sector review 2017.* Nairobi, Kenya, Ministry of Labour and Social Protection, State Department for Social Protection.

Lee, B., Tarimo, K. & Dutta, A. (2018) Tanzania's improved community health fund: An analysis of scale-up plans and design. *Health Policy Plus.* Washington D.C., Palladium.

Mbekeani, K. (2009) Health sector reforms: Tanzania's health policy strategy to increase access to care and improve health. *Medicine and Law: The World Association for Medical Law.* 28 (1), 167–179.

Mills, A., Ally, M., Goudge, J., Gyapong, J. & Mtei, G. (2012) Progress towards universal coverage: The health systems of Ghana, South Africa and Tanzania. *Health Policy and Planning.* 27 (1), 4–12.

Mushi, L. & Millanzi, P. (2019) Health insurance for informal workers: What is hindering uptake? Perspectives from female food vendors in Kinondoni District, Tanzania. *East African Journal of Applied Health Monitoring and Evaluation.* 3, 1–7.

Mwaura, R.N., Barasa, E., Ramana, G.N.V., Coarasa, J. & Rogo, K. (2015) *The path to universal health coverage in Kenya: Repositioning the role of the National Hospital Insurance Fund.* Washington, D.C., World Bank.

NSPF (2008) Ministry of finance and economic affairs, poverty eradication and empowerment division, Dar es Salaam. The United Republic of Tanzania.

NSPP (2003) *National social protection policy.* Tanzania. Available from: http://www.tccia.com/tccia/wp-content/uploads/legal/policy/socialsecuritypolicy.pdf.

NSPP (2011) *Kenya national social protection policy.* Ministry of Gender, Children, and Social Development, Kenya.

NSSF (2018) *National social security fund reports.* Compliance and Collection 2014–2018. Tanzania, National Social Security Fund.

Okungu, V.R. & McIntyre, D. (2019) Does the informal sector in Kenya have financial potential to sustainably prepay for health care? Implications for financing universal health coverage in low-income settings, health systems & reform. *Health Systems & Reform.* 5 (2), 145–157.

Pedersen, R.H. & Jacob, T. (2018) *Social protection in an electorally competitive environment (2): The politics of health insurance in Tanzania: ESID Working Paper 110.* Manchester, UK, The Effective States and Inclusive Development (ESID) Research Centre.

Prabhakaran, S. & Dutta, A. (2017) Actuarial study of the proposed single national health insurance scheme in Tanzania: A summary brief. *Health Policy Plus.* Washington D.C., Palladium.

Riisgaard, L. (2020) *Worker Organisation and Social Protection amongst Informal Petty Traders in Tanzania.* (4 ed.) Roskilde Universitet. CAE Working Paper No. 2020:4.

ROK (2020) *Joint communique: Health sector inter-governmental forum on Universal Health Coverage (UHC), Sarova Whitesands Hotel, Mombasa, October 30, 2020.* Mombasa, Republic of Kenya.

Rwegoshora, H.M.M. (2016) Social security challenges in Tanzania: Transforming the present – protecting the future. Dar es Salaam, Mkuki na Nyota.

SSL (2012) *The social security laws (amendments) act, 5/2012.* International Labour Organization. Geneva, International Labour Office. Available from: https://www.ilo.org/dyn/natlex/natlex4.detail?&p_isn=94062.

Wang, H., Juma, M.A., Rosemberg, N. & Ulisubisya, M.M. (2018) Progressive pathway to universal health coverage in Tanzania: A call for preferential resource allocation targeting the poor. *Health Systems & Reform.* 4 (4), 279–283.

Wanyama, F.O. & McCord, A.G. (2017) *The politics of scaling up social protection in Kenya: ESID Working Paper 87.* Manchester, UK, The Effective States and Inclusive Development (ESID) Research Centre.

Waweru, M. (2017) 12mn people in informal sector yet to register for NHIF. *Capital News.* Available from: https://www.capitalfm.co.ke/news/2017/01/12mn-people-informal-sector-yet-register-nhif/ [Accessed 6 March 2021].

3 The relationship between association membership and access to formal social protection

A cross-sector analysis of informal workers in Kenya and Tanzania

Nina Torm

Introduction and literature

Based on survey data of informal workers gathered across the sectors of construction, trade, and transport, this chapter analyzes the extent to which informal worker associations facilitate access to formal social protection measures in Kenya and Tanzania. Drawing on the Devereux and Sabates-Wheeler (2004) framework, which is outlined in Chapter 1 of this volume, the focus in this chapter is on *preventive* and *promotive*-type social protection measures. *Protective measures*[1] are largely left aside as they fall outside the immediate scope of what informal associations might assist with, and *transformative measures* are looked at in more detail in the sector-specific chapters of this book. The chapter finds inspiration in the Power Resources Approach (PRA) developed by Wright (2000) and Silver (2003) within which the notion of *associational power* is particularly interesting to engage with since it is derived from the formation of collective organizations of workers at various levels.[2] Traditionally applied in the context of the Global North and formal workplaces, such organizations include grass-roots work groups, work councils, and shop-steward bodies as well as trade unions at the industry level (Schmalz, Ludwig & Webster, 2018). In relation to informal workers, associational power may derive from different types of collective associations, institutions, and alliances created or engaged in by workers to advance their own interests and conditions.

For instance, in the case of transport workers in Dar es Salaam, Tanzania, Rizzo (2017) provides a detailed ethnographic account and portrays the rise and decline of a number of worker organizations over a period of almost 20 years. He shows that associations as collective attempts try to address the everyday challenges of daladala workers, such as paying the fines of those charged with traffic offences and supporting burial costs of immediate family. Thus, associations may enable their members to manage the precariousness of their employment (promotive measures) but not to challenge or reform it (through more transformative approaches). Despite the persistence of the necessity for 'agency' and 'voice' to

DOI: 10.4324/9781003173694-3

bring workers together, Rizzo (ibid.) also shows the limitations and vulnerability of worker organizations to assert robust collective action. For instance, challenges arise from the bus owners being small-scale capitalists with little visible movement towards accumulation and growth due to their isolated work situation and low level of skills. In Rizzo (2011), he shows that, by reorganizing, daladala workers in Dar es Salaam created a welfare fund and additional income; yet, "from the outset there existed a structural imbalance between members' needs for support and the association's capacity to provide it" (p. 1196).

Moving to the construction sector in Dar es Salaam, Tanzania, Wells and Jason (2010) find that some of the driving forces and motivations behind the UNDP-funded formation of the umbrella organization TAICO (Tanzanian Association of Informal Construction Workers) in 2006 included the urgent need to increase opportunities for employment and income as well as attaining legal recognition, empowerment, and facilitating access to markets. Amongst some of the achievements, TAICO successfully lobbied the Tanzanian Government to reserve a share of public-sector contracts for informal groups of workers and campaigned for the greater adoption of labour-based construction technologies (ibid.). However, after the UNDP project ended, TAICO collapsed.[3] In the case of Kenya's (informal) construction sector, Mitullah and Wachira (2003) find that umbrella associations, such as the National Federation of Jua Kali Associations of Kenya, which are well-placed for advocacy and policy influence on behalf of informal/small-scale enterprises, have generally performed poorly due to leadership issues. By contrast, KENASVIT has demonstrated more success in organizing and empowering street vendors and informal traders in order to improve their business through training, access to credit, dialogue with local authorities, and other relevant organizations (Mitullah, 2010). Whilst the significance of transnational networks for local struggles is demonstrated through the KENASVIT case, Mitullah (2010) emphasizes the importance of local engagement and capacity-building for sustainable organizing.

Studies of traders from other regions of the Global South also provide positive examples. For instance, focusing on informal street vendors in India, Kumar and Singh (2018) demonstrate the emergence of associational power. In particular, and in response to the growing numbers of street vendors facing harassment, confiscations, and sudden evictions since the 1990s, the National Association of Street Vendors of India (NASVI) was formed as an association of trade unions, community-based organizations, NGOs, and individual members to successfully advocate for street vendors' rights and policy changes. In the transport sector, Spooner and Mwanika (2018) look at how the Amalgamated Transport and General Workers' Union (ATGWU) in Uganda built informal transport workers' associational power through the affiliation of mass-membership associations of informal workers, notably representing minibus taxi workers and motorcycle taxi riders. This strategy of building a hybrid organization has assisted the union in bridging the divide between formal and informal workers, achieving substantial gains for informal workers, and reducing their vulnerability.

Taken together, the studies show that informal self-employed workers, across different sectors, who have low structural power may, with varying degrees of success, create new forms of associational power that diverge from traditional trade union constellations. In addition to member numbers, associational power also depends on infrastructural resources, organizational efficiency, member participation, and internal cohesion, factors which of course vary across different association types and locations. Moreover, associations must optimize their structures so that associational action can be reconciled with the underlying structural conditions and the interests of the members (Schmalz, Ludwig & Webster, 2018). In the current chapter, member interests/needs include different aspects of social protection as captured by preventive and promotive-type measures.

The above brief review shows that existing literature consists mostly of in-depth sector-specific accounts of workers coming together to claim social protection, whilst cross-sector (and cross-country) evidence on such initiatives is scarce. Thus, this chapter provides an empirical contribution to a relatively small but growing literature, by comparing across sectors and locations, the extent to which informal worker associations are able to facilitate member access to formal social protection measures. Additional interpretations and in-depth sector-specific discussions of the extent to which this is related to associational- or other types of power (or the lack thereof) are unwrapped further in the subsequent chapters of the book. The rest of this chapter is structured as follows: the next section briefly introduces the data and method used, along with a summary of selected informal worker characteristics. The subsequent section consists of the analysis and discussion of the main results, followed by a brief conclusion.

Who are the informal workers?

Data and method

This paper uses survey data of informal workers gathered in Tanzania and Kenya during 2018 under the collaborative Danida-funded project as described in Chapter 1. The data was collected for workers engaged in *construction, trade, and transport* – sectors which are all highly prone to informality. Location-wise, data was sampled from two urban areas in each country: Nairobi and Kisumu in Kenya and Dar es Salaam and Dodoma in Tanzania. In each location, three zones/districts were identified through transport hubs, and the same sites were used across the three sectors. The survey covered wage-workers, own-account workers, and micro-businesses with maximum two employees at any given point in time, thus a broad segment of informal worker types. The sampling was done so that 25 percent of workers were sampled through (registered) associations identified purposively for the project. The reasoning behind this was to ensure a broad coverage of different types of associations amongst our respondents, whilst being aware of the potential bias that this introduces to the sample. The remaining 75 percent were sampled by geography to ensure a degree of randomness, bearing in mind that it is not possible to ensure a representative sample when there is no clearly defined population of informal workers (and hence the probability of selection cannot be specified). This random sampling was also intended to capture members of smaller more unestablished worker associations. The initial aim was to interview 1,200 workers in total – 600 per country (100 workers per sector, per site); yet, due to oversampling, the total sample counted 1,462 workers, which, after a thorough cleaning process, resulted in a final sample of 1,385 observations.[4]

Main characteristics

Table 3.1 provides summary statistics for selected variables based on the full data sample *including* the 25 percent of workers that were purposively sampled through associations. In order to check whether the pattern differs for the

Table 3.1 Key summary statistics

	All		*Kenya*		*Tanzania*	
	Mean	*SD*	*Mean*	*SD*	*Mean*	*SD*
Association member*	0.41	0.49	0.48	0.50	0.34	0.48
Formal social insurance enrolment (health/pension)	0.26	0.44	0.34	0.48	0.18	0.39
Health insurance coverage	0.29	0.45	0.41	0.49	0.19	0.39
Nairobi	0.26	0.44	0.57	0.50	0.00	0.00
Kisumu	0.20	0.40	0.43	0.50	0.00	0.00
Dodoma	0.30	0.46	0.00	0.00	0.56	0.50
Dar	0.24	0.42	0.00	0.00	0.44	0.50
Gender (male = 1)	0.77	0.42	0.77	0.42	0.78	0.42
Age	35.23	9.89	36.20	9.80	34.38	9.89
Married	0.66	0.47	0.76	0.43	0.58	0.49
Locally born	0.37	0.48	0.26	0.44	0.46	0.50
Mean daily earnings (current USD)**	10.54	14.22	10.74	9.19	10.36	17.45
Assets (house and/or land)	0.35	0.48	0.27	0.44	0.43	0.50
Primary incomplete	0.10	0.30	0.11	0.32	0.09	0.28
Primary complete	0.53	0.50	0.42	0.49	0.62	0.48
Secondary or above	0.37	0.48	0.47	0.50	0.29	0.45
Professional training course	0.22	0.42	0.27	0.45	0.18	0.39
Training on job	0.25	0.43	0.32	0.47	0.18	0.39
Self-taught	0.53	0.50	0.41	0.49	0.64	0.48
Construction	0.31	0.46	0.34	0.48	0.28	0.45
Trade	0.37	0.48	0.35	0.48	0.39	0.49
Transport	0.32	0.47	0.31	0.46	0.34	0.47
Wage-worker	0.37	0.48	0.55	0.50	0.22	0.41
Own-account	0.52	0.50	0.38	0.49	0.64	0.48
Micro-business	0.11	0.31	0.07	0.26	0.14	0.35
Observations	1,385		644		741	

Source: Author's elaboration based on the survey data.

Notes: Summary statistics based on the full sample. * Association member figures are from the random sample consisting of 979 workers (see also Table A1). ** Median earnings are USD 7.6 for the full sample, USD 6.5 for Tanzania, and USD 8.7 for Kenya.

75 percent of workers sampled randomly, we present the summary statistics for those in Appendix Table A1. With the obvious exception of the *association member* variable, there are no significant differences between the two samples. Thus, the stratification of part of the sample does not seem to impose any bias, at least in terms of the variables chosen, and I therefore proceed using the full sample in the analysis. The association member share (based on the random sample) shows that, overall, 41 percent of workers belong to some type of association with the share being higher in Kenya at 48 percent compared with 34 percent in Tanzania.[5] However, the incidence of association membership varies substantially across the three sectors and by country.[6]

Proceeding to the other variables, Table 3.1 indicates that 26 percent of workers contribute to a formal social insurance (SI) scheme, and the share is almost double in Kenya (34 percent) compared with Tanzania (18 percent).[7] Reassuringly, these figures are in accordance with the official estimates, as indicated in Chapter 2. Across both countries, the vast majority of workers referred to *health insurance* specifically; in fact, out of the 358 workers that answered positively to contributing to some type of SI, 78 percent reported health insurance, 16 percent pension fund, and 7 percent other insurance. The low pension share is not surprising, given that, in Kenya, only about 250,000 of the 14 million informal workers are registered with the National Social Security Fund[8] (NSSF, 2017), with Tanzania having an even lower share, as outlined in Chapter 2. The reasons provided for not being part of any official social insurance scheme include that it is too expensive, and workers have a lack of knowledge about the schemes.[9] In Kenya, for those that do contribute to health insurance, almost 80 percent of workers contribute directly to the provider and 20 percent through an association, compared with the division being 53 percent and 47 percent in Tanzania, a difference which is likely due to the group-based KIKOA scheme in Tanzania, which is explained in more detail in Chapter 2. In both countries, individual National Health Insurance Fund (NHIF) enrolment also covers immediate family members, and thus, we include health insurance coverage as a variable in Table 3.1.[10] Logically, the coverage rate is higher than enrolment, at 29 percent on average and especially for Kenya at 41 percent compared with 19 percent for Tanzania. Reassuringly, these figures are consistent with the national coverage rates estimated at 39 percent in Kenya (KSPSR, 2017) and 22 percent for Tanzania (Jacob & Pedersen, 2018b).

In terms of location, the largest share of workers is based in Dodoma followed by Nairobi, Dar, and Kisumu. The variable "local" indicates that, in Kenya, 26 percent of workers were born in the city where they work; yet, for Tanzania, the share is 46 percent, pointing to a lower incidence of internal migration in the latter, related to a higher level of urbanization in Tanzania compared to Kenya. As for gender, 77 percent of the sample are men, and the mean age is 35 years. Regarding education, 10 percent have below the primary level, whilst 53 percent have completed primary education and 37 percent have completed secondary school. In Kenya, a higher share of workers have secondary education compared with Tanzania; but, in general, workers are well-educated. In terms of training, the vast majority of workers are self-taught at 53 percent overall and as high as 64 percent in Tanzania compared with 41 percent in Kenya. Around 25 percent have had on-the-job training, and 22 percent have attended a professional training school, with both shares being slightly higher in Kenya. In terms of worker types, own-account represent the vast majority with 52 percent on average and as high as 64 percent in Tanzania, whereas, in Kenya, wage-workers make up the largest share with 55 percent, whilst 38 percent are own-account workers.[11] Micro-businesses constitute only around 11 percent of the sample. As for the sectors, the distribution is relatively even, although with a slightly higher share of trade and transport in Tanzania and a somewhat higher share of construction in Kenya.

With regard to the mean daily earnings across all workers, this is around USD 11, only slightly higher in Kenya compared with Tanzania.[12] Interestingly, in the case of Tanzania, earnings are comparable to the 2016 monthly earnings estimate reported for workers in the formal sector of TZS 448,462 (USD 196). When looking at Tanzania earnings in more detail, there seems to be some quite high figures reported, especially in construction, which is also related to the higher share of piece-rate and contractual workers, whereas, in Kenya, the most common form of payment in construction is time-rate.[13] In the case of Kenya, the official monthly earnings estimate from 2017 was KES 57,008 (USD 563), so substantially above the informal earnings from the survey. Thus, the difference in official earnings data between Kenya and Tanzania does not seem to be reflected among informal workers. However, between the three sectors, there is substantial variation in earnings especially in Tanzania, where traders earn well below the formal sector average, whilst construction workers earn above, and transport workers somewhere in the middle. By contrast, in Kenya, the sector-specific earnings distributions are more homogenous.[14] Keeping in mind that informal workers are generally in a less predictable work situation compared with formal workers, this nevertheless seems to suggest that working in the informal construction sector in Tanzania may not always be a last resort option.[15] The large variation in earnings among informal workers in Tanzania is shown by the SD being above the mean in Table 3.1. Thus, for accuracy, we also report the median earnings (in the note of the table) which are well below their respective means due to the right skew of the distribution, whereby the bulk of the sample are relatively low earners, with a few heavy earners at the top pulling up the mean. Due to this skewed earnings distribution in the analysis that follows, the log transformation is used, which makes the distribution appear more symmetric (more normal). Finally, 35 percent of workers own assets, with the share being substantially higher in Tanzania at 43 percent, compared with 25 percent in Kenya. This difference could be related to the finding that Tanzanian workers have a higher likelihood of working where they were born, which means that they are more likely to own property either by inheritance or having bought land (which is generally cheaper in Tanzania compared to Kenya).

Turning to Table 3.2, which displays summary statistics for association members only, in terms of social insurance enrolment, the share is 30 percent, which is

Table 3.2 Key worker characteristics, association members

	All		*Kenya*		*Tanzania*	
	Mean	*SD*	*Mean*	*SD*	*Mean*	*SD*
Formal social insurance enrolment (health/ pension)	0.30	0.46	0.40	0.49	0.21	0.41
Health insurance coverage	0.34	0.47	0.47	0.50	0.22	0.41
Association type						
Sacco/vicoba/chama	0.51	0.50	0.64	0.48	0.39	0.49
Work-related association	0.44	0.50	0.29	0.45	0.58	0.49
Women/youth/religious	0.05	0.22	0.07	0.25	0.04	0.19
Benefit type						
Work-related	0.37	0.48	0.34	0.47	0.40	0.49
Social cushioning	0.13	0.33	0.05	0.21	0.20	0.40
Voice and representation	0.05	0.21	0.04	0.20	0.05	0.22
Loans	0.46	0.50	0.57	0.50	0.35	0.48
Barriers (yes = 1)	0.64	0.48	0.62	0.49	0.66	0.48
Association fee (yes = 1)	0.73	0.44	0.77	0.42	0.69	0.46
Observations	800		388		412	

Source: Author's elaboration based on the survey data.

higher than that for all workers (Table 3.1), and again, enrolment incidence is the double in Kenya compared with Tanzania. As expected, health insurance coverage is also higher for members at 34 percent (compared with 29 percent for all), and for Kenya, the 47 percent coverage is quite a bit above the national average. As for the various types of associations, more than 50 percent report belonging to a sacco/vicoba/chama, which is obviously a rather broad categorization that may encapsulate a variety of types of associations and services/functions. As with social insurance enrolment, there is also substantial variation by country, in that the incidence of sacco/vicoba/chama is significantly higher in Kenya at 64 percent compared with 39 percent in Tanzania. Overall, 44 percent belong to a worker association, and here, the pattern is the opposite with a higher share in Tanzania at 58 percent compared with 29 percent in Kenya. Finally, around 5 percent of workers belong to either women or youth-specific or other types of association.

Another way of categorizing associations is to look at the different kinds of benefits that they provide, and here, 46 percent of workers report that the main benefit is the opportunity to save and receive loans, whilst 37 percent state that it is work-related.[16] Another 13 percent answer that the main benefit is social cushioning, which includes access to formal health (and pension) insurance as well as informal support for children's education, funerals, weddings, etc.[17] Finally, 5 percent answer protection against harassment, eviction, or confiscation of goods, which may be interpreted as voice and representation. As for the types of associations, the distribution of variables varies somewhat by country with a higher than average work-related and social cushioning share in Tanzania, and a higher than average loans share in Kenya. Finally, the majority of respondents (64 percent) answer that there are barriers to enter associations and 73 percent report paying an association fee, the latter being higher in Kenya (77 percent) compared with Tanzania (69 percent).[18] Such differences in terms of associations' services and access across countries and between sectors are explored further in the sector-specific chapters of this volume.[19]

Association members versus non-members

Since this paper looks at the extent to which association membership is associated with formal SI, it is of interest to look at any significant differences across the relevant variables when comparing members with non-members. First, as expected, Table 3.3 shows that association members are significantly more likely to be enrolled in a social insurance scheme, which, as indicated earlier, covers mostly health-related protection. In fact, 30 percent of members contribute to social insurance compared to 20 percent of non-members. When health insurance coverage is considered, the shares are 34 percent and 22 percent, respectively. As for

Table 3.3 Differences in key workers characteristics by association member status

	Member	*Not Member*	*Difference*	*t-Value*
Formal social insurance enrolment (health/pension)	0.30	0.20	0.11	4.53***
Health insurance coverage	0.34	0.22	0.12	4.85
Nairobi	0.26	0.26	0.00	0.02
Kisumu	0.22	0.17	0.05	2.15**
Dodoma	0.30	0.29	0.01	0.51
Dar	0.21	0.27	-0.06	-2.61**
Gender (male = 1)	0.72	0.84	-0.12	-5.19***
Married	0.72	0.59	0.12	4.83***
Local	0.39	0.35	0.04	1.48
Age	36.56	33.40	3.16	5.95***
Mean daily earnings (current USD)*	10.84	10.48	0.36	0.46
Assets (house and/or land)	0.38	0.32	0.06	2.21**
Primary incomplete	0.09	0.11	-0.02	-1.30
Primary complete	0.54	0.51	0.03	1.26
Secondary or above	0.37	0.38	-0.01	-0.50
Training course	0.25	0.19	0.05	2.21**
Training on job	0.23	0.26	-0.03	-1.31
Self-taught	0.52	0.54	-0.02	-0.71
Construction	0.24	0.40	-0.17	-6.71***
Trade	0.38	0.35	0.02	0.90
Transport	0.39	0.24	0.14	5.64***
Wage-worker	0.34	0.42	-0.08	-3.15***
Own-account	0.53	0.51	0.02	0.62
Micro-business	0.14	0.07	0.06	3.82***
Observations	1,385			

Source: Author's elaboration based on the survey data.

locations, Kisumu workers are more likely to be association members, whereas, in Dar, workers are more likely *not* to be members. There is a higher share of men amongst non-members than among members, which matches the fact that construction workers are significantly more likely to be non-members.[20] By contrast, in the transport sector, workers are more likely to be association members, whereas, in trade, there is no significant difference. Association members are older, more likely to be married and have assets, and they are also more likely to have attended a professional training course, but, otherwise, educational background does not differ by membership. Finally, micro-businesses are more likely to be association members and the opposite for wage-workers, whilst for own-account workers – the vast majority – the distribution is more or less equal.

Table 3.3 reveals that earnings do not differ significantly between members and non-members; yet, interestingly, when looking only at Kenya, association members have higher average incomes than non-members, whereas, in Tanzania, the opposite is the case.[21] Recalling from Table 3.2 that Kenyan workers have a relatively higher membership share in associations where receiving loans is reported as the main benefit, it is perhaps not surprising that these workers, on average, have higher earnings than non-association members. On the other hand, Tanzanian workers have a higher share of members in worker associations that report social cushioning and work-related benefits as being the most important. Following on, from this, I turn to look at what distinguishes workers that are enrolled in formal protection schemes from those that are not.

Social insurance enrolment versus non-enrolment

In Table 3.3, we saw that there was a relation between being association member and contributing to formal SI, and this is supported by Table 3.4 which shows that 68 percent of those enrolled in formal SI are also association members, whereas members constitute a lower share of those that are *not* enrolled (54 percent).[22] In terms of the types of associations, there is a significantly higher share of those workers in SI enrolment that are sacco/vicoba/chama members compared with those that are not enrolled. Correspondingly, enrolled workers are more likely to be receiving loans from their associations, possess assets, and have higher earnings compared with those not enrolled. SI receivers are older than non-receivers, and they are also more likely to be married. Nairobi has the largest share of workers enrolled in formal SI schemes, whereas Dar es Salaam has a higher proportion of non-enrolled workers, compared to Dodoma which has a relatively high share of enrolled workers, likely due to the community health fund scheme. In terms of education, those enrolled are significantly better educated, and non-enrollers are more likely to be self-taught. As for the sectors, only transport workers differ by being more represented amongst those enrolled and the same for wage-workers and micro-businesses, whilst own-account workers have a higher likelihood of not contributing to SI. Having uncovered significant differences by association membership status and SI enrolment along a variety of worker characteristics, we now turn to examining the extent to which being

Table 3.4 Differences in key workers characteristics by social insurance enrolment

	Social Insurance	*No Social Insurance*	*Difference*	*t-Value*
Association member	0.68	0.54	0.14	4.53***
Sacco/vicoba/chama	0.40	0.26	0.14	5.20***
Work-related association	0.24	0.26	-0.02	-0.59
Women/youth/religious	0.04	0.03	0.01	0.77
Work-related	0.24	0.21	0.03	1.27
Loans	0.35	0.23	0.12	4.54***
Voice and representation	0.03	0.03	0.00	0.17
Social cushioning	0.06	0.08	-0.02	-1.03
Nairobi	0.37	0.22	0.15	5.58***
Kisumu	0.25	0.19	0.06	2.43**
Dodoma	0.28	0.31	-0.03	-0.97
Dar	0.10	0.28	-0.18	-7.10***
Gender (male = 1)	0.78	0.77	0.01	0.32
Married	0.79	0.62	0.17	6.08***
Local	0.38	0.37	0.01	0.27
Age	37.05	34.59	2.46	4.08***
Mean daily earnings (current USD)	12.77	9.96	2.80	3.19***
Assets (house and/or land)	0.40	0.34	0.06	1.94*
Primary incomplete	0.06	0.11	-0.05	-2.97**
Primary complete	0.44	0.56	-0.11	-3.76***
Secondary or above	0.50	0.33	0.17	5.76***
Training course	0.32	0.19	0.12	4.88***
Training on job	0.25	0.24	0.00	0.16
Self-taught	0.44	0.56	-0.13	-4.20***
Construction	0.29	0.31	-0.03	-0.95
Trade	0.33	0.38	-0.05	-1.73
Transport	0.38	0.30	0.08	2.72**
Wage-worker	0.44	0.35	0.09	3.08***
Own-account	0.41	0.56	-0.15	-5.06***
Micro-business	0.16	0.09	0.06	3.35***
Observations	1,385			

Source: Author's elaboration based on the survey data.

member of an informal worker association is related to participation in a formal social insurance scheme.

Association membership, earnings, and access to formal social security

In order to examine the correlation between being member of an informal worker association and access to formal SI, I use a standard probit model, where SI is the outcome (dependent) variable taking a binary form and the same for the main variable of interest which is association membership.[23] Table 3.5 column 1 reports the correlation between association membership and contributing to SI when key worker characteristics are controlled for. The results show that members are 27 percent more likely to contribute to SI compared with non-members,

Table 3.5 Association membership and formal social insurance

	(1)	*(2)*	*(3)*
Member	0.265*** (0.080)	0.198** (0.083)	
Male	-0.105 (0.098)	-0.049 (0.120)	-0.029 (0.122)
Age	-0.010 (0.024)	-0.001 (0.024)	-0.000 (0.024)
Age^2	0.021 (0.029)	0.012 (0.030)	0.011 (0.030)
Married	0.357*** (0.090)	0.295*** (0.093)	0.307*** (0.093)
Assets	0.018 (0.080)	0.108 (0.084)	0.111 (0.084)
Local	0.113 (0.079)	0.145* (0.086)	0.153* (0.087)
Secondary education+	0.426*** (0.121)	0.471*** (0.123)	0.471*** (0.123)
Training course	0.168 (0.105)	0.095 (0.110)	0.096 (0.110)
Training on job	0.036 (0.097)	-0.009 (0.103)	-0.009 (0.103)
Mean daily earnings (USD)	0.076* (0.046)	0.056 (0.048)	0.055 (0.048)
Micro-business	0.318*** (0.122)	0.372*** (0.126)	0.386*** (0.126)
Wage-worker	0.274*** (0.089)	0.214** (0.106)	0.240** (0.107)
Nairobi		0.801*** (0.125)	0.747*** (0.128)
Kisumu		0.546*** (0.131)	0.498*** (0.134)
Dodoma		0.444*** (0.123)	0.435*** (0.124)
Trade		0.189 (0.137)	0.177 (0.138)
Transport		0.307*** (0.105)	0.314*** (0.105)
Work-related			0.156 (0.105)
Social cushioning			-0.013 (0.161)
Voice and representation			0.059 (0.241)
Loans			0.318*** (0.099)
N	1,385	1,385	1,385

Source: Author's elaboration based on the survey data.
Notes: Marginal effects for the probit estimations. Dep. variable: worker contributes to formal health/social insurance. Robust standard errors in parentheses. *p < 0.10, **p < 0.05, ***p < 0.01.

ceteris paribus. In column 2, location and sector dummies are added which reduces the likelihood of association members participating in SI to 20 percent. In terms of the various control variables, I note that married and more educated workers that are classified as wage-workers or micro-businesses in the transport sector (as compared with construction and micro-trade) are more likely to contribute to SI. The reason that transport workers are significantly more likely to participate in SI is due to the specific set-up for informal transport workers in Kenya, whereby matatu workers are legally required to belong to a Sacco/transport management company, and these may commission certain savings to pay for the workers' NHIF (and NSSF) coverage. In terms of location, workers in Nairobi, Kisumu, or Dodoma have a higher chance of being part of a formal SI scheme compared with those based in Dar. Across locations, individuals that are working in the city in which they were also born have a higher chance of formal SI enrolment which is probably related to them having more knowledge of different schemes. The significance of the earnings variable in column 1 disappears when the location and sector dummies are added in column 2, indicating that earnings vary by location and/or sector, as discussed previously.[24]

In column 3, the membership variable is split into different types of associations as proxied by the key benefits they provide, revealing that the loan-type associations (whether saccos/vicobas/chamas or other informal groups) are driving the results.[25] Since one of the key functions of Saccos is loan provision either directly or indirectly via establishing agreements with banks, one interpretation could be that members use the loans to pay for SI contributions. Yet, that would suggest that SI participation is also determined by earnings or assets, which Table 3.5 shows is not the case as neither determine SI participation when other factors are taken into account. The variable for social cushioning-type associations is not well-determined; yet, the negative sign indicates that workers already receiving social support (*preventative* social protection measures) from their associations are less likely to be in a formal SI scheme, as would be expected.[26] Members of loan-type associations (*promotive* social protection measures) are substantially (around 32 percent) more likely to have access to SI compared with non-members. Hence, these kinds of associations are likely to serve a dual purpose of (a) ensuring some sort of financial cushioning for their members and (b) facilitating access to public insurance schemes. In relation to (a), one might then expect to find a correlation between association membership and earnings, which I examine further on, whilst (b) seems to be the result of a particular type of associational setup or purpose, potentially suggesting the presence of a degree of 'associational power'.

For instance, many of the bodaboda workers in Kenya belong to self-governed motorcycle associations like the Bodaboda Association of Kenya (BAK) which is characterized by strong associational power and works to further workers interests:

> Many members have NHIF at individual level as compared to NSSF. BAK is working to have the members make contribution daily so that they cater for hospital bills, funeral expenses and in case of accident to the rider, the family members can continue earning something. There are plans to link members to insurance schemes. There is already an insurance agency in place.
>
> (KII, Nairobi)

As for micro-traders, the data reveal that around half of the workers have come to know about the NHIF scheme from their social network, including their associations (mostly *chamas*), and some of these play a crucial role in terms of registering members with the NHIF:

> ...All members of my group are registered for NHIF and are contributing KES 200 every month through the group. Every week, the group takes a sum of KES 7000 to the NHIF. If a member is unable to pay the KES 200 the group takes the money from the members' savings and pays for him or her. The member who is being paid for is considered to have a debt of the group which he or she must pay.
>
> (FGD, Kisumu)

Thus, in addition to facilitating enrolment, this particular group also prevents default among their members, although this is not common practice and generally SI payments are made individually. In the case of construction, where participation in formal social insurance schemes is less common, an association member in Kenya points out:

> People are different. Not all members of the group have NHIF. There are those that are being pushed to have insurance. We live from hand to mouth and it also makes it hard for us to pay the NHIF.
>
> (FGD, Nairobi)

For the particular association referred to here, members have NHIF on an individual basis (or as a family), and the members mentioned that it is very difficult to have insurance as a group. However, the association is encouraging members to have NHIF.

In Tanzania, the link between membership and formal SI relates partly to the "KIKOA" scheme which under the NHIF Mutual Plan provided health packages tailored to groups of informal workers with 20 or more members. Due to the relatively affordable premium, different types of worker associations (including saccos, vicobas, bodaboda, and daladala group) were motivated to register, in turn, receiving SI for their members. The scheme began in 2015, yet was terminated at the end of 2019 due to issues of adverse selection (see Chapter 2 in this volume for more detail).

In combination with the survey data, such insights explain the observed positive relation between membership and SI, whilst pointing to sector (and location)-specific differences which are explored further in the other chapters of the book.

Turning to the earnings analysis, Table 3.6 presents the OLS results when the dependent variable is the log of mean earnings (in USD), and again, our main variable of interest is association membership. Following a standard Mincer earnings function (Mincer, 1974), we control for education and training of the worker and use worker age as a proxy for experience.[27] In addition, we add job function, sector, and location as these are also expected to account for some of the variation in earnings. In column 1, when worker-specific controls are included, association members have earnings that are 12 percent higher compared with non-members, and in column 2, when location and sector dummies are added, they fall slightly to 11.6 percent. As for the controls, men earn between 36 percent and 49 percent more than women, which is higher than in much of the gender pay gap literature.[28] Earnings increase with age (albeit at a diminishing rate) and assets, whereas workers living in the city they were born have lower earnings. In line with the literature, workers that have a post-secondary education and have undergone training have higher earnings (see Bjerge, Torm & Trifković, 2021). Wage-workers earn substantially below own-account workers, who, in turn, earn less than micro-businesses, and workers in Kenya have higher earnings compared with Dar, whilst Dodoma is at the bottom.

As for the different sectors, transport workers earn more than micro-traders; yet, construction workers are the highest earners on average (especially

Table 3.6 Association membership and earnings

	(1)	(2)	(3)
Member	0.125*** (0.047)	0.116** (0.046)	
Male	0.485*** (0.066)	0.355*** (0.074)	0.358*** (0.074)
Age	0.059*** (0.015)	0.048*** (0.014)	0.048*** (0.014)
Age^2	-0.075*** (0.019)	-0.064*** (0.018)	-0.064*** (0.018)
Married	0.055 (0.055)	0.004 (0.055)	0.005 (0.055)
Assets	0.058 (0.049)	0.114** (0.050)	0.115** (0.050)
Local	-0.193*** (0.049)	-0.107** (0.053)	-0.106** (0.053)
Secondary education+	0.293*** (0.074)	0.256*** (0.075)	0.256*** (0.075)
Training course	0.275*** (0.068)	0.183*** (0.071)	0.183*** (0.071)
Training on job	0.244*** (0.059)	0.124** (0.060)	0.125** (0.060)
Micro-firm	0.215** (0.086)	0.231*** (0.084)	0.233*** (0.085)
Wage-worker	-0.169*** (0.052)	-0.403*** (0.057)	-0.399*** (0.058)
Nairobi		0.309*** (0.065)	0.302*** (0.066)
Kisumu		0.260*** (0.075)	0.251*** (0.075)
Dodoma		-0.135* (0.070)	-0.139** (0.070)
Construction		0.129** (0.054)	0.128** (0.054)
Trade		-0.301*** (0.071)	-0.305*** (0.071)
Work-related			0.093 (0.058)
Social cushioning			0.113 (0.080)
Voice and representation			0.103 (0.143)
Loans			0.137** (0.059)
Constant	0.329 (0.284)	0.753*** (0.279)	0.755*** (0.279)
N	1,385	1,385	1,385

Source: Author's elaboration based on the survey data.
Notes: Marginal effects for the OLS estimations. Dep. variable: worker earnings (log mean in USD). Robust standard errors in parentheses. *p < 0.10, **p < 0.05, ***p < 0.01.

in Tanzania). In column 3, the association membership variable is again split into the association/benefit types revealing that the membership "effect" is driven by workers in loan-type associations. This indeed suggests that one mechanism through which loan-type associations enable their members' access to public insurance schemes is through the provision of loans. However, it might also be that high earning workers chose to be association members and are able to pay for social insurance regardless of membership, using the loans for expanding their business, etc. (in turn, leading to even higher earnings). In addition to loans, work-related benefit-type associations are close to being well-determined and associated with higher earnings, which is not surprising given that this category includes a more secure job contract and higher or more stable income, etc.

The general finding that members earn more than non-members is supported by one of the KII from the transport sector in Kenya, which reveals that Sacco members, in this case, earn more than workers in the formal sector:

> We want the boda-boda job to be respected as we have seen that even the kind of money we get is even more than that of those in formal employment,

> some of our members are getting more than even teachers in the government employment.
>
> (KII, Kisumu)

When splitting by country, differences across the two countries are revealed in terms of the kinds of workers associations attract and what workers use their associations for. For instance, in Kenya, associations mostly facilitate access to formal systems of health insurance (preventive measures) and play a promotive role such as providing micro-credit/loans, as also discussed in other chapters of the book. By contrast, in Tanzania, workers seem to use associations more as cushioning mechanisms and for other more transformative types of social protection protection (e.g. voice and representation).

In the survey, earnings are reported as current and so is membership status; thus, there is a possibility that the observed positive relation between association membership and earnings is driven by reverse causality if high earners self-select into associations rather than higher earnings being an outcome of membership. This possibility is investigated further in Torm (2020), and the preliminary findings indicate that earnings are indeed an outcome of association membership and not vice versa.

To sum up, the results show that association members, in particular, those that are members of loan-type associations are substantially more likely to be enrolled in formal SI schemes. Second, and relatedly, association members have earnings that are significantly higher than non-members, when other relevant factors are controlled for. These findings hold in general, yet the underlying mechanism through which social protection coverage occurs via associations differs by sector and country, as does the extent to which this is explained by associational power or other types of power resources. These relations are explored further in other chapters of this edited book.

Conclusion

Against the background of an increasing global drive towards universal social protection, this paper has set out to analyze the extent to which informal workers, through association membership, are able to access formal social protection schemes in urban areas of Kenya and Tanzania. In both countries, existing formal SI schemes are open to informal workers, yet uptake is generally limited due to various factors including lack of information and limited efforts in reaching out to informal workers. In this context, informal worker associations, in a variety of constellations, may act to bridge the divide, in addition to potentially providing direct social support to their members. For instance, initiatives like saving clubs and representation of members towards authorities and/or employers also fall within the broader conceptualization of social protection, acknowledging the fact that informal worker associations often are engaged in numerous activities.

In order to understand the motivations behind workers coming together in associations, this chapter draws on the PRA, the premise of which is that organized labour can successfully defend its interests by collective mobilization of power, be it ***structural**, **associational**, **symbolic**, **logistical**, **or institutional power**.* In the current chapter, ***associational power*** provides a particularly interesting analytical concept as it emerges from different types of collective organizations, institutions, and alliances created or engaged in by informal workers to advance their own interests and conditions. In fact, work from across the Global South has shown that informal workers have, in some cases, been quite successful in creating new forms of ***associational power***, which diverge from traditional power constellations.

Whereas previous studies mostly consist of sector-specific in-depth cases void of a comparative aspect, the current contribution makes use of cross-sectoral survey data from four urban locations in Kenya and Tanzania, thus adding new insight on informal workers and the outcomes of associational belonging. The results reveal that association members are significantly more likely to access formal social insurance, controlling for key worker characteristics. Moreover, across sectors and sites, the provision of loans appears to be the main direct mechanism through which this occurs. In addition, the analysis reveals a substantial earnings-gap between association members and non-members, and this gap seems to vary substantially from Kenya to Tanzania. This suggests that workers may use associations for different purposes ranging from basic welfare cushioning to preventive, promotive, and/or more transformative types of social protection, which is discussed further in the other chapters of the book. Overall, the findings suggest a potentially important role for informal worker initiatives both in terms of providing direct social cushioning and indirectly through enabling participation in existing SI schemes.

Appendix A

Table A1 Summary statistics, random sample

	All		*Kenya*		*Tanzania*	
	Mean	*SD*	*Mean*	*SD*	*Mean*	*SD*
Member	0.41	0.49	0.48	0.50	0.34	0.48
Social protection enrolment	0.24	0.43	0.31	0.46	0.17	0.38
Health insurance coverage	0.26	0.44	0.37	0.48	0.16	0.37
Nairobi	0.28	0.45	0.57	0.49	0.00	0.00
Kisumu	0.21	0.41	0.43	0.49	0.00	0.00
Dodoma	0.26	0.44	0.00	0.00	0.52	0.50
Dar	0.24	0.43	0.00	0.00	0.48	0.50
Gender (male = 1)	0.80	0.40	0.78	0.42	0.82	0.38
Age	34.02	9.72	35.24	10.07	32.81	9.23
Married	0.64	0.48	0.73	0.45	0.55	0.50
Local	0.36	0.48	0.25	0.43	0.46	0.50
Mean daily earnings (USD)	10.36	14.15	10.18	8.44	10.53	18.10
Assets	0.33	0.47	0.27	0.44	0.39	0.49
Primary incomplete	0.10	0.30	0.11	0.31	0.09	0.29
Primary complete	0.52	0.50	0.41	0.49	0.62	0.49
Secondary or above	0.38	0.49	0.48	0.50	0.29	0.45
Training course	0.21	0.41	0.26	0.44	0.16	0.37
Training on job	0.24	0.43	0.30	0.46	0.18	0.39
Self-taught	0.55	0.50	0.44	0.50	0.66	0.48
Construction	0.33	0.47	0.35	0.48	0.31	0.46
Trade	0.35	0.48	0.34	0.47	0.36	0.48
Transport	0.32	0.47	0.31	0.46	0.33	0.47
Wage-worker	0.40	0.49	0.56	0.50	0.25	0.44
Own-account	0.50	0.50	0.39	0.49	0.62	0.49
Micro-business	0.09	0.29	0.05	0.23	0.13	0.34
Observations	979		486		493	

Table A2 Summary statistics, association members only, random sample

	All		*Kenya*		*Tanzania*	
	Mean	*SD*	*Mean*	*SD*	*Mean*	*SD*
Social protection enrolment	0.30	0.46	0.37	0.49	0.20	0.40
Health insurance coverage	0.32	0.47	0.42	0.49	0.18	0.39
Sacco/vicoba/chama	0.63	0.48	0.80	0.40	0.41	0.49
Work-related association	0.29	0.46	0.13	0.33	0.52	0.50
Other	0.07	0.26	0.08	0.27	0.06	0.25
Work-related	0.33	0.47	0.30	0.46	0.38	0.49
Social cushioning	0.12	0.33	0.03	0.17	0.25	0.44
Voice and representation	0.03	0.18	0.04	0.19	0.03	0.17
Loans	0.51	0.50	0.63	0.48	0.34	0.48
Barriers (yes = 1)	0.56	0.50	0.51	0.50	0.61	0.49
Association fee (yes = 1)	0.68	0.47	0.70	0.46	0.66	0.47
Observations	405		235		170	

Appendix B

a Sampling method

Throughout the book, I adopt a broad definition of informal employment, including own-account/self-employment in informal enterprises (i.e. unregistered micro-business) as well as wage employment in informal jobs (i.e. without a written contract but possibly working for a formally registered enterprise). The broader term "informal economy", which also encompasses agricultural employment (not covered in this volume), was endorsed by the International Labour Conference (ILC) in 2002 (ILO, 2002; Chen, 2005) and is now commonly used instead of the older and narrower concept of the informal sector. In accordance with this broader definition of the informal economy, the survey targeted wage-workers, own-account workers, and micro-business owners with a maximum of two employees. For wage employment in informal jobs and in relation to contracts, I asked potential interviews to specify first whether they had a contract, if so whether it specifies pay and entitlement to benefits, and if so whether the details of the contract are implemented in practice. If the answer to the first two were yes, but the answer to the last one was no, I proceed with the interview on the basis of this being an informal worker.

In terms of the sectors, first, for transport, the aim was to divide the sub-sample equally between bodaboda and daladala/matatu and to sample both drivers and conductors and riders. For micro-traders, the target group were traders that were mobile (on the street, at bus terminals, vacant lots, etc.) and also less mobile ones, for example, mama lishe, but excluding those with permanent structures such as kiosks or regular designated markets. In addition, the enumerators were instructed not to over cover some commodities but rather to mix different types of traders, different types of commodities without distinguishing between what workers sold. For construction, the target groups were skilled and unskilled workers (wage-workers and own-account) employed directly by construction/site managers or indirectly via an intermediary, for example, gang-leaders. As for the construction sites, I covered large and medium construction/building sites, waiting sites, for example, streets, buildings, and excluded individual residential housing sites.

The aim was to cover 1,200 workers in total (600 per country) 200 workers per sector (100 per sector per site), where 25 percent were sampled through associations and 75 percent by geography. For those workers sampled from associations, the target group were ordinary members and not leaders or members in an official position. Associations covered formal associations/SACCOs/trade unions (i.e. registered associations). For the geographical/random part of the sample, in each country, two urban locations were selected (Kenya: Nairobi and Kisumu; Tanzania: Dar es Salaam and Dodoma) and three zones/districts per site were identified through transport hubs. The same sites were used across the three sectors. In Tanzania, the construction sector sampling was also partly based on snowballing techniques in order to identify large and medium construction/building sites. The data gathering began in June 2018, and in the first phase, 75 percent of the targeted workers were surveyed (by geography) using the Survey-to-Go software hosted by the Institute for Development Studies (IDS) of the University of Nairobi. Subsequently, information on associations that informal sector workers belonged to was obtained and extracted in preparation for the second survey data collection phase which was undertaken in November and December 2018. This phase targeted 25 percent of workers drawn from the associations covered in phase one of data collection.

b Cleaning the data

After merging the two datasets from Kenya and Tanzania, I had a total sample of 1,462 workers, yet were left with a final sample of 1,385 observations after a couple of cleaning steps as listed below:

i Removing workers that were not in the target group, including tuk-tuk drivers and bicycles;
ii Dropping workers that did not provide earnings or provided earnings above the 99th and below the 1st percentile (outliers) by country and sector;
iii Dropping workers that indicated association contributions above the 99th and below the first percentile (outliers) by country and sector.

Notes

1 Protective measures in the form of cash transfers are usually state-level interventions. See, for instance, Jacob and Pedersen (2018a) for a case study of productive social safety nets in the case of Tanzania. For Kenya, the NSPP (2011) provides a summary of formal social assistance programmes going back to 2010. Moreover, Hickey and Seekings (2017) provide a historical and conceptual account of the promotion of cash transfers in Sub-Saharan Africa, in general.
2 See Chapter 1 for a detailed description of the PRA.
3 Talks of TAICO becoming a member of the official Tanzania Mines and Construction Workers Union (TAMICO) stalled for a number of reasons including TAMICO's reluctance to welcoming members with irregular incomes and changing employers (Wells & Jason, 2010).
4 For further detail on the sampling strategy and cleaning procedure, refer to Appendix B.
5 The full sample obviously has a higher membership share with 58 percent overall and 60 percent/56 percent for Kenya and Tanzania, respectively.

6 In Kenya, the association membership share is 33 percent, 47 percent, and 57 percent for construction, trade, and transport, respectively, and for Tanzania, the equivalent figures are 19 percent, 34 percent, and 50 percent.

7 The question asked was "Do you contribute to an insurance scheme, for example, NHIF, national pension fund, or another insurance?"

8 NSSF presentation by Nancy Mwangi on 'practical perspective of the NSSF scheme' (Nairobi, 2017).

9 In Tanzania, out of the 80 percent that do not contribute any official social insurance scheme, 24 percent say it is too expensive; 23 percent say they have no knowledge about insurance and 12 percent say that they have no time/were never given NHIF registration. In Kenya, the major reason is that it is too expensive reported by 30 percent followed by 9 percent saying that they have no time/were never given NHIF registration. Hence, a lack of knowledge about insurance schemes is more pre-dominant in Tanzania than in Kenya. Across the countries, of the 74 percent that do not contribute around half are association members. Out of those workers that contribute, in terms of health insurance specifically, 38 percent in Kenya indicate that they receive this from a public insurance scheme, whilst 21 percent get it from a private provider. In Tanzania, the vast majority (66 percent) get it from a public insurance scheme, whilst 6 percent get it from a private provider.

10 Coverage is defined as when either the worker him/herself, the spouse, children, or another relative is the primary NHIF contributor. In the vast majority of cases and across both countries, the contribution comes mostly from the worker him/herself.

11 This discrepancy between the two countries relates partly to the groupings of especially construction workers which more often have been classified as wage-workers in Kenya and own-account workers in Tanzania (the latter is in line with the Tanzanian 2000/2001 Integrated Labor Force Survey (ILFS) where 60 percent of construction workers in Tanzania are self-employed).

12 Daily earnings are converted into the sample median of six days a week and 11 hours per day. Reported working hours are slightly longer in Tanzania than Kenya. Throughout the chapter, the average USD rate from June 2018 to December 2018 is used (101.2 for Kenya and 2,284.5 for Tanzania, per USD).

13 According to the Tanzania ILFS (NBS, 2014), (formal) workers in the construction and transport sector earn around 320,000 TZS monthly (USD 140).

14 Interestingly, in Kenya, construction is the sector with the lowest earnings and trade the highest, whilst it is the reverse in Tanzania. In relation to construction, this can be explained by the fact that the Tanzania sample includes more masons, who are ranked higher compared with the helpers (wage workers hired by the masons) who constituted the majority of the Kenya construction sample.

15 This is also supported by 19 percent of Tanzania workers responding "money being good" to the question of why they chose their job whilst 16 percent respond increased income/profit and 49 percent answer able to care for my family. In Kenya, as many as 31 percent respond the "money being good" and 78 percent respond that the main benefit of their current job is financial benefits/income.

16 Work-related benefits include, for instance, a more secure job contract, higher or more stable income, enhanced information that helps to earn more in the job, safety, better information about alternative job offers, access to services, etc. Those workers that report loans as the main association benefit also report receiving bereavement payment. Important also to note that what workers report as the main benefit of association membership is most often not the sole benefit, as the majority of associations have multiples purposes/offer a range of services.

17 Note that the access to formal health (and pension) insurance component included in the category of social cushioning represents a rather small share of 3 percent across both countries (slightly higher for Kenya). If workers self-select into associations to gain access to for instance health insurance, then the correlation between association

membership and official insurance schemes would be biased. Thus, for the purpose of the forthcoming analysis, it is reassuring that workers do not appear to select into associations with the purpose of gaining access to official social security schemes.

18 As with Table 3.1, the random sample equivalent is presented in Appendix A Table A2, showing that the share of SACCO members is higher whilst the worker association representation is lower, especially in Kenya. Correspondingly, the benefit types show that the incidence of loans is higher in the random sample, whereas the work-related measure is lower. This difference is the result of the purpose-based sampling of 25 percent of workers allowing for a broader representation of different types of associations, even if the random landscape looks different.

19 Further analysis using multinomial regressions to examine the challenges that informal workers face and the key benefits they receive from their associations reveal that both challenges and benefits are quite sector and country-specific and also vary substantially by worker-types. Interestingly, however, neither challenges nor benefits differ much by gender, income, education, or age of the workers.

20 Nevertheless, 23 percent of the members are construction workers, which although a lower share than the other two sectors seems to be a rise compared with Mitullah and Wachira (2003) which in the case of Nairobi found that most construction workers did not belong to any association.

21 The difference in wages between members and non-members is statistically significant in the case of Kenya as revealed when performing t-tests by membership status split by country location, whereas, in Tanzania, the difference is not statistically significant (results not reported).

22 Reassuringly, no significant differences emerge when health insurance coverage is used instead of social protection enrolment.

23 When health insurance coverage is used as the outcome instead of SI, the results remain qualitatively very similar.

24 As recalled from Table 3.4, workers in formal SI enrolment have significantly higher earnings; yet, Table 3.5 shows that earnings do not seem to be a determinant of social security participation, when other factors are taken into account.

25 Where workers have classified that the main benefit is receiving loans, this could represent either a sacco/vicoba/chama-type association or a work-related association, but the defining characteristics is that the association provides loans.

26 The fact that the work-related and voice and representation-type associations do not show any statistical significance in terms of enabling SI access for their members does not exclude the possibility that these association types provide important social protection measures and act as potential mechanisms for workers to voice concern. For instance, in Tanzania, 43 percent of association members have contacted their associations on issues related to representation, and 40 percent of workers say that this helped resolve the issue. In Kenya, 30 percent of workers have made contact on issues related to representation, and out of those, 30 percent say that it helped resolve the issue.

27 I include age squared to allow for a diminishing marginal effect regarding the earnings returns to experience.

28 I also note that the gender pay gap is substantially higher in Tanzania (results not reported).

References

Bjerge, B., Torm, N. & Trifković, N. (2021) Can training close the gender wage gap? Evidence from Vietnamese SMEs. *Oxford Development* Studies. 49 (2), 119–132.

Chen, M.A. (2005) *Rethinking the informal economy: Linkages with the formal economy and the formal regulatory environment.* Research Paper 2005/010. Helsinki, UNU-WIDER.

Devereux, S. & Sabates-Wheeler, R. (2004) *Transformative social protection.* IDS Working Paper 232. Brighton, Institute of Development Studies.

Hickey, S. & Seekings, J. (2017) *The global politics of social protection.* WIDER Working Paper 2017/115. Helsinki, UNU-WIDER.

ILO (2002) *Report VI, Decent work and the informal economy*, Sixth item on the agenda. In: International Labour Organization. *International Labour Conference 90th Session.* Geneva, International Labour Office.

Jacob, T. & Pedersen, R.H. (2018a) *Social protection in an electorally competitive environment (1): The politics of productive social safety nets (PSSN) in Tanzania: ESID Working Paper 109.* Manchester, UK, The Effective States and Inclusive Development (ESID) Research Centre.

Jacob, T. & Pedersen, R.H. (2018b) *Social protection in an electorally competitive environment (2): The politics of health insurance in Tanzania: ESID Working Paper 110.* Manchester, UK, The Effective States and Inclusive Development (ESID) Research Centre.

KSPSR (2017) *Kenya social protection sector review 2017.* Ministry of Labour and Social Protection, State Department for Social Protection. Nairobi, Kenya.

Kumar, S. & Singh, A.K. (2018) Securing, leveraging and sustaining power for street vendors in India. *Global Labour Journal.* 9 (2), 135–149.

NSPP (2011) *Kenya national social protection policy.* Ministry of Gender, Children, and Social Development, Kenya.

NSSF (2017) Presentation on 'practical perspective of the NSSF scheme' Workshop Report: Social Protection and Informal Economy International Launch Workshop (Nairobi, 2017).

Mincer, J. (1974) *Schooling, experience, and earnings.* National Bureau of Economic Research.

Mitullah, W.V. (2010) Informal workers in Kenya and transnational organizing: Networking and leveraging resources. In: Lindell, I. (ed.) *Africa's informal workers: Collective agency, alliances and transnational organizing in urban Africa.* London, Zed Publications, pp. 184–203.

Mitullah, W.V. & Wachira, I.N. (2003) *Informal labour in the construction industry in Kenya: A case study of Nairobi.* Geneva, International Labour Office.

NBS (2014) *Integrated labour force survey.* Dar es Salaam, National Bureau of Statistics.

Rizzo, M. (2011) 'Life is war': Informal transport workers and neoliberalism in Tanzania 1998–2009. *Development and Change.* 42 (5), 1179–1206.

Rizzo, M. (2017) *Taken for a ride: Grounding neoliberalism, precarious labour, and public transport in an African metropolis.* Oxford, Oxford University Press.

Schmalz, S., Ludwig, C. & Webster, E. (2018) The power resources approach: Developments and challenges. *Global Labour Journal.* 9 (2), 113–134.

Silver, B.J. (2003) *Forces of labor: Workers' movements and globalization since 1870.* Cambridge, Cambridge University Press.

Spooner, D. & Mwanika, J.M. (2018) Transforming transport unions through mass organisation of informal workers: A case study of the ATGWU in Uganda. *Global Labour Journal.* 9 (2), 150–166.

Torm, N. (2020). Social protection and the role of informal worker associations: A cross-sector analysis of urban sites in Kenya and Tanzania. *Roskilde University.* CAE Working Paper Nr. 2020:3.

Wells, J. & Jason, A. (2010) Employment relationships and organizing strategies in the informal construction sector. *African Studies Quarterly.* 11 (2&3), 107–123.

Wright, E.O. (2000) Working-class power, capitalist-class interests, and class compromise. *American Journal of Sociology.* 105 (4), 957–1002.

4 Self-regulating informal transport workers and the quest for social protection in Tanzania

Godbertha Kinyondo

Introduction

In Tanzania, 90 percent of the cars, trucks, and buses are privately owned, and most people rely on affordable informal sector transportation such as motorcycles (bodaboda), rickshaws (bajaji), and minibuses (daladala).[1] These informal transportation modes provide an important source of livelihood for operators (Cervero, 2000). The transportation sector accounts for 6 percent of total informal sector employment in the country and is male-dominated (ILFS, 2014; NBS, 2014a). In 2014/2015, the transportation and storage industry employed 2.6 percent of the 20 million working population in Tanzania. Workers in the transportation sector work up to 14 hours per day almost seven days a week, which violates labour laws. The Employment and Labour Law Act of 2004 (URT, 2004; PDL, 2014) stipulates that an employee may work six days a week, nine hours per day, or a total of 45 hours in a week.

The informal transport sector is prone to injuries and fatalities. In November 2018, the Deputy Minister for Home Affairs reported 8,237 deaths and 37,521 cases of lost-limb injuries involving motorcycle accidents between 2008 and 2018 (Vanguard, 2018). From 2016 to 2018, 3,719 motorcycle riders were injured in 4,624 road accidents (AllAfrica, 2019). Amidst such high levels of risks, the majority have no access to social health insurance. Often, informal transportation workers are the main family breadwinners, and their deaths and lifetime injuries can be a cause of poverty for the family (Mbegu & Mjema, 2019).

Poor access to Social protection for informal transport workers mirrors the inadequate provision of social protection for the general population. Tanzania spends only 2 percent of its GDP on income security needs of older people, one of the lowest ratios in the world (ILO, 2018). In terms of health, in 2018, only 34 percent of the population was covered by health insurance. The largest insurance schemes are the National Health Insurance Fund (NHIF) and the Community Health Fund (CHF). NHIF is mandatory for civil servants and covers only 7 percent of the population (Wang et al., 2018; UNICEF, 2018). CHF, which is a public scheme meant to serve the rural and poor, covers 25 percent of the population, and private insurance covers 2 percent. In 2018, Tanzania merged the five duplicative pension schemes into two schemes: Public Service Social Security

DOI: 10.4324/9781003173694-4

Fund (PSSSF) and the National Social Security Fund (NSSF). The new PSSSF serves the public sector[2] while the NSSF caters for the informal sector[3] and private-sector workers. Significantly, informal transport workers rely on their associations for their social protection.

Against the backdrop of the inadequacy of social protection for informal workers in Tanzania, this chapter employs the Devereux and Sabates-Wheeler (2004) social protection framework and utilizes the Power Resource Approach (PRA) to assess the relevance of transportation associations in meeting the social protection needs of daladala and bodaboda operators. It explores worker characteristics, social protection opportunities and challenges, forms of organizing, and how informal transport associations assist members in acquiring social protection. The chapter draws on survey data, focus group discussions (FGD), and Key informant Interviews (KII) conducted between August 2018 and February 2019 in Dar es Salaam and Dodoma. The survey covered 250 bodaboda and daladala drivers and conductors. The KII included association leaders and stakeholders in the governance of the informal transport sector, while FGDs included only members of selected associations.

The chapter is divided into nine sections. The following section looks at governance of the transport sector. The third section gives three examples of transportation associations followed by an analysis in section four based on the PRA. The fifth section presents the profile of the surveyed informal transportation workers, and section six looks at the key work-related threats and challenges facing informal transport workers. Section seven utilizes the Devereux and Sabates-Wheeler (2004) framework to describe the measures that informal associations take in regard to social protection and section eight discusses pertinent issues related to informal transport workers, their associations and SP, while section nine concludes and makes recommendations.

Governance of the transportation sector

This section describes how, in the face of the government's inability to effectively govern the informal transportation sector, informal sector workers take it upon themselves to coordinate their own operations.

Government transportation institutions, policies, and regulations

The Ministry of Works, Transport and Communications (MOWTC) is responsible for establishing transportation policy and has a broad mandate of overall planning and investment in roads, rail, waterways, and marine and air corridors. The National Transportation Policy of 2003 "allows private sector participation and market competition" but also calls for government oversight and regulation so that "workers and customers are protected against abuse" (URT, 2003). While MOWTC collects statistics related to motorcycles, bajaji, and minibuses, it does not exercise direct oversight of the informal transport sector. The Tanzanian National Roads Agency (TANROADS), a semi-autonomous agency within MOWTC, is responsible for the maintenance and development of trunk and regional road networks and undertakes road safety audits.

The Land Transport Regulatory Authority (LATRA) was established by the Land Transport Regulatory Authority Act No. 3 of 2019 to regulate movement of goods and passengers by commuter buses, intercity buses, goods-carrying vehicles, taxi, motorcycles, and tricycles. LATRA replaced the Surface and Marine Transport Regulatory Authority (SUMATRA). In collaboration with the Dar es Salaam Commuter Bus Owners Association (DARCOBOA), LATRA allocates routes and fee structures for buses moving between urban areas.

The Ministry of Labour and Ministry of Health's are responsible for policies, that is, Social Security and Health Sector Strategic Plan, respectively, which envisage social protection for all but, unfortunately, lacks an implementation plan for the informal sector.

At the local level, the Local government authorities (LGAs) are responsible for the development and management of the public transport infrastructure like roads, terminals, and bus stops. LGAs issue vehicle licenses for bodaboda and daladala and establish passenger collection points and parking areas. The Police issue driving licenses and are responsible for enforcing safe driving behaviour and ensuring roadworthiness of vehicles. Further, they provide safety and security training for transportation workers and conduct rider awareness campaigns through print and broadcast media.

Effectiveness of government transportation institutions

Because LATRA does not have the broad scope of SUMATRA, it is hoped that its transportation oversight will be more effective. For example, at one point, SUMATRA required daladala owners to show legal contracts between them and licensed drivers before their mini-buses could be registered to operate for business. The frequency of submitting manipulated documents was so high that SUMATRA eventually abandoned the requirement. Also, SUMATRA sometimes assigned routes that had few customers. Drivers were thus tempted to drive on non-designed routes or cut the trip short. SUMATRA's defence was that it had insufficient staff and means for conducting adequate review and surveillance. Hopefully, LATRA will not have the same constraints.

LATRA's delegation of authority to LGAs to issue vehicle licenses and to set up parking and passenger collection points is meant to inspire local solutions to problems and to enhance enforcement of regulations and surveillance of bodaboda and daladala behaviour. However, LGAs are often stretched for resources and capacity to conduct these functions. Also, they may unwisely use their authority. For instance, the Regional Commissioner (RC) of Dar es Salaam ordered daladala to transport teachers without payment of bus fare on top of onboarding school children at a concessionary rate. This caused conductors to fill buses quickly with other passengers, drivers to skip pickup points where there were many students, and teachers to hide their IDs until after taking their seats. This unnecessary conflict and undue economic pressure could have been avoided by consulting local stakeholders first.

Bodaboda disreputably move in and out of traffic lanes in disregard of traffic regulations and respect for other road users. Police rarely stop them for crossing intersections on red lights because it is a common behaviour. Bodaboda are keenly aware of the threat of having an accident due to their reckless behaviour, but this does not deter them from engaging in risky driving to accumulate adequate money for the day. Many passengers refuse to wear helmets because they are considered uncomfortable and dirty, and again enforcement is lax.

This section indicates fragmented strategies, lacking vertical and horizontal coordination, amongst echelons of government agencies, regarding the need, requirement, and regulation of the informal transport sector. Furthermore, agencies have limited financial and human resources to relate to informal transportation workers.

Self-governance and collective organizing among informal transportation workers

Over the years, daladala and bodaboda groups have arisen progressively, bound together by the social obligation to assist each other in need or under threat. They assemble in parking areas and around passenger collection points where they naturally share common problems in dealing with authorities or employers, as noted below:

> our group came together to assist ourselves when one of us get involved in an accident, death, sickness of himself and or close family members; harassment by authority, conflict with bosses; we have a constitution that explain member's benefits depending on the situation...
>
> (FGD participant DSM)

One of the common strategies of informal transport groups to gain negotiation leverage is to associate with a formal union. For example, Umoja wa Madereva na Dar es Salaam (UWAMADAR) was established by 42 drivers and conductors on 12 July 1997. In order to have an audience with the authorities, they were compelled to be affiliated with a trade union, COTWU-T. COTWU-T requested that UWAMADAR increase its membership to be a more legitimate entity, and by 2003, the association had over 5,236 members (Rizzo, 2020).

However, UWAMADAR subsequently was weakened by leaders, who failed to call regular meetings because they, like their members, were on the road most of the time as fulltime drivers. Furthermore, UWAMADAR did not meet the requirements of legal representation, which is why it became an affiliate of COTWU-T. Eventually, UWAMADAR leaders left to form the Tanzania Road Transport Workers Union (TAROTWU), which represents drivers, conductors, and others working on buses. COTWU-T faced sector-specific factions[4] as well, that is, Umoja wa Madereva Tanzania (UWAMATA), which represents only drivers who operate on upcountry routes; CHAWAMATA, representing drivers of lorries, and Tanzania Association of Drivers Union (TADU) that represents bodaboda workers and those that operate smaller vehicles like bajaji.

Participation in transportation associations carries some risk. Elected officers, who may not be versed in financial management, can be duped. In some instances, a leader, who has personal needs or ambitions, may embezzle funds, which may go undetected for a long period. These leadership problems often result in the dissolution of associations and later re-establishment with a new leadership.

Generally, most informal transportation associations lack sufficient and durable power to negotiate with the authorities on their own. Furthermore, the constant fragmentation and disintegration of their associations only strengthen the hand of the authorities, who will continue to favour negotiating with unions which represent the formal transportation sector.

Despite internal weakness and external constraints under which informal transportation associations operate, several have demonstrated that it is possible for bodaboda and daladala to organize and become effective representative bodies. As the three examples below show, enlightened and able leadership is one of the key ingredients of success.

Communications Transportation Workers Union of Tanzania[5]

COTWU-T is an affiliate of the Trade Union Congress of Tanzania (TUCTA), which is the apex trade union in the country. It has nearly 6,000 members from the transport and telecommunication sectors. It operates through democratic structures of elected members to executive committees at national and branch levels and conducts a quadrennial conference. It has a woman's officer on its national executive committee.

The union mainly organizes formally employed workers in the transport and communication sector. However, faced with dwindling membership and encouraged by TUCTA, the union created an informal economy desk (IEDO) for the purpose of organizing informal transport workers. The IEDO participates in international fora to discuss key issues facing informal transport workers (KII, IEDO, Secretary General). TUCTA developed a draft policy that would mobilize the informal sector to form associations that would, in turn, link to larger associations and affiliate unions (TUCTA 2015).

COTWU-T has been proactive in reaching out to the informal transportation sector. They routinely invite leaders of informal transportation groups to their seminars and workshops. Facilitators include representatives from national government, such as NHIF, NSSF, and the police, which gives association leaders opportunity to not only learn but also nurture critical contacts at higher levels. These trainings also address matters such as social security, health insurance, and other support schemes.

When COTWU-T learned that the Matangini Michungwani King'ongo (MMK) association with 250 bodaboda members was having difficulties in registering as a formal association, COTWU-T intervened, which resulted in MMK gaining legal status as an affiliate branch of COTWU-T in April 2019. Prior to that, MMK existed as an informal self-help group of workers with the aim of welfare supporting one another. Despite COTWU-T reaching out to informal transportation worker groups, MMK has been one of the few associations taking advantage of what COTWU-T offers. On several occasions, COTWU-T has made legal services available to MMK, mostly disputes between bodaboda and the owners of motorcycles. MMK members credit COTWU-T for the membership ID card which protects them from harassment by authorities (FGD with MMK, 2019). MMK members of COTWU-T pay a monthly fee of TZS 1,000 or any amount dimmable, while formal members pay 2 percent of their salaries. MMK gained prominence after COTWU-T invited them to participate in the 'Labour Day parade', and their placard read: "Bodaboda is like any other job".

Chama cha Madereva wa PikiPiki Dar-es-Salaam[6]

Chama cha Madereva wa PikiPiki Dar-es-Salaam (CMPD) is an umbrella association of bodaboda drivers, boasting a total of 45,000 members in Dar es Salaam and was registered formally in September 2016. The association was established as a result of a meeting in April 2016 with the RC for Dar es Salaam. The RC, a young activist political leader, listened to their grievances about strenuous working conditions, abuse by owners, and harassment by traffic police. This resulted in forming a committee that included representatives of SUMATRA, police, and officials from the five districts.

The RC's request was that bodabodas form one single group with which he and the committee could continue discussions. At the time, there were three registered district-level bodaboda associations: from Ilala District, Chama Cha Waendesha Pikipiki Ilala (CHAWAPILA), from Kinondoni District, Chama Cha Bodaboda Wilaya ya Kinondoni (CHABOWIKI), and Temeke District, Umoja Wa Waendesha Pikipiki Temeke (UWAPITE). There were also various informal groups in Kigamboni and Ubungo districts.

With this high-level political support, CMPD negotiated with district authorities to reduce the area of the city centre that was prohibited entry by bodaboda. CMPD have since mediated several disputes between bodaboda drivers and motorcycle owners, who are more open to discussion, knowing that CMPD enjoys political support. CMPD also mediate disputes between bodaboda, daladala, and private drivers over access to different parking areas and over harassment of bodaboda as noted by a bodaboda respondent, "Drivers of big vehicles contribute towards bodaboda accidents as they do not respect bodaboda drivers when driving. They sometimes push bodaboda drivers deliberately to compel them to clear the way".

CMPD conducts trainings on safety and health issues and credits itself for reducing accidents and drivers' deaths. Other trainings have included topics on businesses management and customer service. The association has intervened to help members get driving licenses. CMPD established a SACCO for bodaboda drivers, which, in turn, has enabled an agreement with Equity Bank to provide loans to help members purchase their own motorbikes (conversation with CMPD Deputy Secretary, 2019).

CMPD succeeded in recruiting three women who are bodaboda operators. Given dearth of women in the sector, this has greatly enhanced the reputation of the association. A less-favourable reputational aspect occurred in 2019 when the secretary of the association was accused of embezzlement of funds and was suspended from the association pending court hearing. However, this has not endangered the association, which is currently being guided by a popular and progressive leader.

Umoja wa Madereva na Makondakta Tegeta Nyuki, Bunju

Umoja wa Madereva na Makondakta Tegeta Nyuki, Bunju (UWAMATEBU) began as an informal group of daladala drivers and conductors in 2006 and was registered in 2014 in Kinondoni District of Dar es Salaam with 100 members. The group was formed in response to abuse by police and passengers and to support members injured in road crashes and funeral expenses. The association has a Board of Directors, which includes a leader from the local government of Kinondoni.

UWAMATEBU has purchased eight daladala registered to the association. This allows the association to provide temporary employment to members who find themselves with no jobs. They are expected to seek employment on other daladala within a specified period. In addition, UWAMATEBU helps conductors who lose their jobs to find temporary jobs as 'wapiga debe' ('drum beater') who help direct passengers to daladala and as fees collectors.

The constitution of the association establishes different levels of social insurance (SI). For example, TZS 100,000 (USD 43.78) is provided to support members, who have death in the immediate family. Upon death in a family, it is culturally required to leave work to care for a surviving parent, parent-in-law, or other close relative during the mourning period. Similarly, the constitution stipulates payment for those who fall ill or are involved in a crash. These funds include profits earned by the eight UWAMATEBU buses and the fees collected at bus stands.

The association has registered with the NHIF so that members can purchase health plans. However, only few members have been able to make payments to acquire the health insurance cards. UMAWATEBU has entered into agreements with local banks to give positive consideration in providing loans to their members and the association itself. In August 2017 and 2018, the association received loans from Akiba Commercial Bank (ACB) worth TZS 50 million and TZS 80 million, respectively, using as collateral land title deeds of the Chairman of the association and the local political supporter on the Board. The loans were used to purchase more buses for the association.

The key to the success of UWAMATEBU has been unusually open-minded male leaders. They actively recruited eight women conductors to become members. One of them is attending driving school, partially at the association's expense. The association promotes the notion that women are skilled individuals who should be proud to be working in daladala. These messages of worth convey upon members a strong sense of agency.

The RC of Dar es Salaam and a local politician recognized UWAMATEBU's work and facilitated their promotion from being a route association to a daladala association representing the District of Kinondoni. The association is currently known as 'Umoja wa Madereva Wilaya ya Kinondoni' (UWAMAWIKI).[7] UWAMAWIKI is registered at the District Executive Directorate and is responsible for putting in place an orderly queuing system, collection of daily dues, and welfare schemes on a district basis. Furthermore, through training, UWAMAWIKI make members aware of the laws and regulations pertaining to vehicle operation and nurture pride in occupation, which has always been looked down upon. The improved discipline enhanced the image of their members and enabled UWAMAWIKI to acquire a tender from the District to transport over 5,000 visitors from SADC countries, who gathered in Dar es Salaam for the SADC Industrialization Week, 5–9 August 2019. Although UWAMAWIKI is not fully institutionalized within the District structures,[8] there exists a rule-based governance system, which, among others, enables smooth collection of government taxes.

The power resources of informal transportation associations

This section borrows from the power resources approach (PRA), which is founded on the basic premise that the workforce can successfully defend its interests by collective mobilization of power. The three examples above illustrate how informal transportation associations demonstrate varying degrees of associational, societal, and institutional power (Schmalz, Ludwig & Webster, 2018). In comparison with the vehicle owners and business interests, drivers and conductors are generally at a disadvantage with respect to power resources, but through their capacity for collective action, the extent of their disadvantage can vary over time.

Associational power

According to Schmalz, Ludwig, and Webster (2018), associational power arises from the action of workers uniting to create collective political power as an entity such as trade union or as a worker association. The strength of associational power is associated with a number of factors such as effective organizational structure, number of members, member willingness to take action such as participating in campaigns, willingness to pay dues, and a strong collective identity.

Daladala and bodaboda groups are organized into different types of associations, whether registered or not registered with authorities. Some are organized for the purposes of mediating with police, for example, when a member runs afoul of traffic regulations or when they collectively meet with authorities to establish parking spaces. Many other associations are created mainly for the purposes of social welfare and protection.

Most informal transport associations are not able to attract a large and stable membership. This is particularly true of bodaboda, who are in constant search of areas where there are more customers. Daladala drivers and conductors can be dismissed by their employers and are forced to go to other areas of the city. Sometimes, members cannot pay membership fees regularly as required, which limits resources available to the association leaders to network and advocate on behalf of the members. The fragmentation caused by groups breaking away from existing association to form their own association and the lack of able or trustworthy leaders further reduces the power of associations. However, information gathered through FGDs and the survey reveals that members valued being part of their small associations, as noted by members:

> … in case of unfair dismissal by owner of vehicles, leaders of association accommodate one as "DAY WORKER" and contacts other associations till you find a vehicle.
>
> (Daladala respondent, Dodoma)

> …We are small, our leaders come from within us, with low education, they don't know much, and government does not listen to us…but these associations assist us to collect money when someone is sick or have an accident, but least in big issues that need solutions.
>
> (Daladala respondent, Dar es Salaam)

Societal power

Schmalz, Ludwig, and Webster (2018) define societal power as society's support for worker demands (i.e. discursive power) and cooperation between associations and other social groups and organizations (i.e. coalitional power). An example of discursive power in Tanzania occurred on May Day in 2019 when a group of bodaboda and COTWU-T marched in Dar es Salaam with banners that read: "A Bodaboda job is like any other job". This was reported positively by the media with citizens giving testimony on how much they relied on bodaboda. This got the attention of the Dar es Salaam RC, and the Mayor of the city and bodaboda representatives were invited to their respective offices for consultation regarding their work. This "May Day" event also demonstrated coalition power of COWTU-T extending its reputation and influence to the bodaboda drivers. This had spill-over effects with police offering to provide safety and security training, banks willing to discuss provision of loans, and the NHIF making presentations on health insurance packages.

Institutional power

Schmalz, Ludwig, and Webster (2018) refer to institutional power as legally fixed rights of organized labour. At the national level, this is found in the constitution and legislation. The constitution of the United Republic of Tanzania of 1977 establishes the right of association. The National Employment Promotion Act of 1999 gives a seat to a representative of an informal association on the National Employment Advisory Committee. This legislation was in response to massive retrenchments in the 1990s under Structural Adjustment Programs (SAPs) and the explosive growth of the informal sector, which became the refugee home of many former civil servants. There is little evidence of the committee currently being active on informal sector matters. According to the Labour Officer, the Act is currently being amended. The Employment and Labour Relations Act of 2004 under Sub-Part D: Freedom of Association gives employees the right to form and join a trade union (URT, 2004; ILO, 2004). The Act makes provision for amalgamation of registered trade unions. It prohibits the amalgamation of an unregistered unions with a registered union. Informal associations could affiliate with a registered union to gain access to its institutional power.

The Tanzania Union of Industrial and Commercial Workers (TUICO) is a notable example of a trade union extending its institutional power to the informal transportation sector. TUICO has formed coalition with bodaboda that park along the Mchikichini market, near Kariakoo in Dar es Salaam.

In addition, although short-lived, the coalition with COTWU-T enabled UWAMADAR to access COTWU-T's institutional power, and MMK is currently a branch of COTWU-T leaping on its institution power, such as representation.

The two independent transportation unions, TAROTWU, TADU, and UWAMATA (association), recognizing that they have insufficient institutional power as separate entities, are attempting to form a federation of informal transportation workers. It is uncertain as to whether their federation will attain sufficient institutional power without the support of the formal sector federations, most notably TUCTA. Unions lobby for formalization via a legal requirement that would bring 'dispersed ownership' of vehicles under a 'company', thus establishing employer–employee relationship, hence legal rights, and rights to social protection.

Apart from trade unions, some associations of informal transport workers – like in the example of UWAMAWIKI above – do participate in district-level rule-based governance system related to land planning and taxation issues even if these are not formally institutionalized.

Structural power

The existence of many registered and unregistered informal transportation groups is well-known to public officials, and it is recognized that they pose a formidable mass-membership that could create considerable disturbance if provoked.

Structural power derives from the status of workers in their relationship to the economy. According to Rizzo & Atzeni (2020), workers in the informal transportation sector in Tanzania have potential workplace structural power because of their large numbers and how widely their services are used. Bodaboda have been so hard to regulate because they "possess a certain collective power related to their large number and the service they provide: when they go on strike, urban transport is thrown into chaos since most of the city population depend on them" (Goodfellow & Titeca, 2012, p.14). However, the structure of the economy, with oversupply of available labour, coupled with ease of entry requirements in the sector erodes workers' structural power as they compete in the marketplace for employment.

The competition for jobs even creates mistrust among those already working in the informal transportation sector. This was conveyed poignantly by one of the drivers of daladala in Dodoma:

> ...with this type of work, you learn not to trust, even your co-workers, because our jobs are not guaranteed, your co-worker knowing that he might lose his job, he can go to your "Boss" and inform him that you are not working hard enough, and that if he gets the vehicle he can bring in more money than you do, ... then abruptly, you are terminated, and the co-worker takes over the vehicle—this is common practise in our line of work.

Profile of informal transportation sector respondents

Based on the random sample survey, 50 percent of informal transport workers are association members, with the share being substantially higher in Dar es Salaam compared with Dodoma (Table 4.1). As for individual SI enrolment, the share is 17 percent overall and about double in Dodoma compared with Dar es

Table 4.1 Key worker characteristics

	All		*Dar*		*Dodoma*	
	Mean	*SD*	*Mean*	*SD*	*Mean*	*SD*
Association member*	0.50	0.50	0.54	0.50	0.46	0.50
Social insurance enrolment	0.17	0.37	0.11	0.31	0.21	0.41
Health insurance coverage	0.16	0.36	0.07	0.25	0.22	0.41
Gender (male = 1)	0.99	0.11	0.97	0.17	1.00	0.00
Age	32.21	8.16	33.50	8.75	31.32	7.62
Married	0.52	0.50	0.53	0.50	0.51	0.50
Locally born	0.51	0.50	0.31	0.47	0.64	0.48
Mean daily earnings (current USD)**	9.86	16.45	11.71	19.26	8.59	14.12
Assets (house and/or land)	0.39	0.49	0.37	0.49	0.41	0.49
Primary incomplete	0.04	0.21	0.01	0.10	0.07	0.25
Primary complete	0.66	0.48	0.70	0.46	0.63	0.48
Secondary or above	0.30	0.46	0.29	0.46	0.30	0.46
Professional training course	0.25	0.43	0.23	0.42	0.26	0.44
Training on job	0.16	0.36	0.21	0.41	0.12	0.33
Self-taught	0.60	0.49	0.57	0.50	0.61	0.49
Wage-worker	0.32	0.47	0.27	0.45	0.34	0.48
Own-account	0.60	0.49	0.60	0.49	0.60	0.49
Micro-business	0.08	0.28	0.13	0.34	0.05	0.23
Observations	250		102		148	

Source: Author's elaboration based on the project survey data.
Notes: * Association member figures are from the random sample. **Median earnings are USD 7 for the full sample, 7.7 USD for Dar, and USD 6.6 for Dodoma.

Salaam. The health insurance coverage is also substantially higher in Dodoma, which is likely due to the availability of CHF, currently known as iCHF (improved CHF). The sector is young and male-dominated (99 percent) with an average age of 32 years, whilst around 50 percent are married. Nearly one-third (31 percent) and two-thirds (64 percent) were born in Dar es Salaam and Dodoma, respectively, a notable difference between these groups. The high migration rate among Dar es Salaam respondents mirrors the survey of ILFS (2014), which recorded 55.9 percent of residents of Dar es Salaam having migrated from other parts of the country to earn a living in a commercial city.

Mean daily earnings are around 10 USD and somewhat higher in Dar es Salaam compared with Dodoma, and this geographical difference is also reflected among bodaboda operators who negotiate rates directly with their customers.[9] Dar es Salaam bodaboda earned an average of TZS 37,758 (USD 16.53) per day, while the average earning of Dodoma bodaboda was substantially lower at TZS 19,274 (USD 8.44) per day. As indicated in the footnote of Table 4.1, the median wage is USD 7 which is lower than the mean, as heavy earners at the top of the wage distribution pull up the mean. About 40 percent of the sample of workers owned a plot or a house with the share being slightly higher in Dodoma.

All respondents had some education with two-thirds (66 percent) having completed primary school and 30 percent having completed secondary school (Form IV), across locations. The relatively high level of education among transport workers was noted by a police constable in Dodoma who commented:

> Given the dearth of employment for the youth, parents have opted to buy motorcycles for their Secondary School graduates to use them for business purposes, mostly, while waiting to go for further studies. But most of them end up making it their full-time jobs...

One Form IV graduate confirmed this, "... I have been here for about 3 years... when I graduated, I anticipated getting an office jobs, that dream had long vanished; bodaboda is my bread and butter".

In terms of further training, around 60 percent of the workers were self-taught,[10] while 25 percent had attended a professional training course, and 16 percent had done training on the job, with the latter making up a substantially higher share in Dar. In terms of worker types, whilst a significant proportion were wage-workers (32 percent), with the share being higher in Dodoma (34 percent), the majority (60 percent) of both the Dar es Salaam and Dodoma respondents were own-account workers, and 8 percent considered themselves as micro-businesses, slightly higher in Dar. Most of the own-account workers were bodaboda (86.8 percent), with only few daladala drivers (10.3 percent) and conductors (2.9 percent) indicating own-account status, even though they do not own vehicles. The reasoning provided by these respondents is very revealing. "We are the ones who bring in the money for the bosses, therefore, we are the business men and not workers" (FGD, Dodoma). Another offered:

> ...look here, we only hire the Boss's car to do 'our business', we handle all the logistics, make negotiations with traffic officers, and handle vehicle's technical problems for the day, while hassling to obtain enough money for the Boss, and for ourselves; without extra work and effort, you go empty handed at home.
>
> (FGD, Dar es Salaam)

Whereas daladala workers do not generally own the vehicles, a sizable number of bodaboda own their motorcycles partially due to government initiatives (i.e. LGA Act, 1982, amended 2018)[11] and often have the option of leasing the motorcycle for a specified time, usually a year, after which the bodaboda owns the motorcycle.

Looking at association members only, Table 4.2 shows that SI enrolment and health insurance coverage are only slightly higher than for the full sample, indicating that membership to associations is not a panacea to enhance formal social protection. In terms of the different types of associations, the majority (77 percent) belong to worker-type associations, followed by 22 percent being in a Saccos and Vicoba/chama. Whilst the incidence of the latter is almost double in Dodoma, the worker associations share is higher in Dar.

In Dar es Salaam, in order to access resources, training, and services such as loans, transport workers are required to form associations, and these are also encouraged as a means to facilitate taxation and self-regulation of the industry including identification of individual workers. Workers in Dodoma prefer Saccos and Vicoba as they avail funds for school fees and enable access to capital for business and daily assistance to meet consumption.

Table 4.2 Key worker characteristics, association members

	All		*Dar*		*Dodoma*	
	Mean	*SD*	*Mean*	*SD*	*Mean*	*SD*
Social insurance enrolment	0.18	0.39	0.15	0.36	0.21	0.41
Health insurance coverage	0.17	0.37	0.09	0.29	0.22	0.42
Association type						
Sacco/vicoba/chama	0.22	0.42	0.15	0.36	0.27	0.45
Worker association	0.77	0.42	0.82	0.39	0.73	0.45
Other association	0.01	0.11	0.03	0.17	0.00	0.00
Benefit type						
Work-related	0.42	0.50	0.52	0.50	0.35	0.48
Social cushioning	0.26	0.44	0.18	0.39	0.31	0.47
Voice and representation	0.07	0.25	0.12	0.33	0.03	0.17
Loans	0.25	0.44	0.18	0.39	0.30	0.46
Barriers	0.80	0.40	0.79	0.41	0.80	0.40
Fee	0.68	0.47	0.69	0.47	0.68	0.47
Observations	163		67		96	

Source: Author's elaboration based on the project survey data.

Key work-related threats and challenges

Harassment/arrest/corruption

The survey data (see Figure 4.1) and qualitative information gathered in FGDs indicate that the greatest perceived threat faced by the workers is harassment by the authorities, most notably traffic police, who look for problems and might create them in order to extract bribes. This induces some drivers to violate traffic rules as noted below:

> You have work stress, thinking of earning sufficient money, then the police stop you for no apparent reason. We, Bodaboda, have learnt to move on, even if its red lights, if police stop you, run, otherwise, they'll create a false claim to exert money from you. Imagine, this goes on the whole day.
>
> (FGD-bodaboda, DSM)

Bodaboda particularly complain of harassment by the special police task force commonly referred to as 'Tigo'[12] who monitor bodaboda in areas that the normal traffic officers do not frequent. They do not wear uniforms, move about on motorcycles, and search for thieves and criminals, including bodaboda engaged in criminal activities.

> The Police are our worst enemy. They have what they call 'TIGO' who are there to catch us, this is not fair. Imagine – it's hard enough to get money for oneself, submission to boss, and then pay bribes to police.
>
> (FGD, Bodaboda – Dar es Salaam)

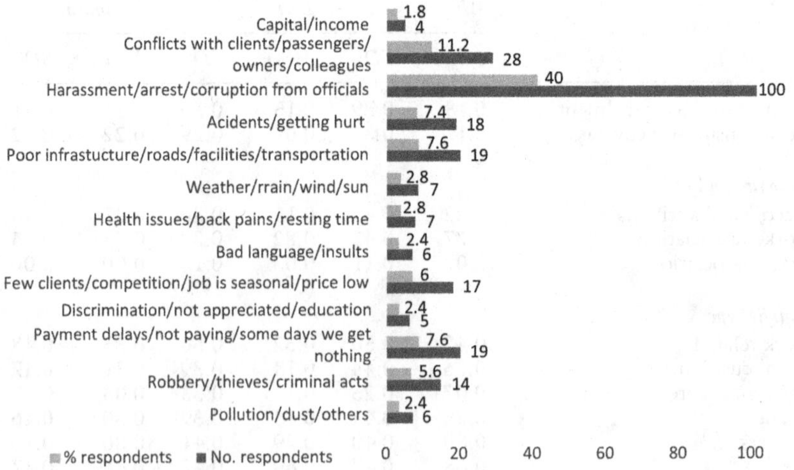

Figure 4.1 Challenges and threats facing informal transportation workers.

Harassment allegations by authorities are documented (Cervero, 2000; Rizzo, 2011; SID, 2012; James, 2014), but police accuse the informal transport workers of failure to adhere to rules and regulations.

Conflicts with clients/passengers/colleagues

As seen in Figure 4.1, getting into conflict with passengers is the second most cited challenge. Some passengers complain of poor language, reckless driving, etc., but some mistreat workers, especially women conductors, for example, some men will refuse to pay the full fare or will flirt with them.

At times, the drivers will intervene to support their female conductors. Women also face threats of sexual harassment by drivers, who demand sexual favours before they can get and retain jobs as conductors.

> …as women we face many challenges coming from unwanted sexual harassment. I have been forced to move almost five times [for refusing] drivers who wanted me…I find myself being kicked out, and sometimes they don't inform you, but you come to work, you find the driver has another conductor. There is nothing you can do; therefore, you go to look for another job.
>
> (FGD, Daladala, DSM)

The prevailing perception of women obtaining jobs as conductors through sexual favours presents a challenge to well-intended drivers.

> …many drivers would like to hire more women conductors, because women are trusted with the day collection. However, it's a problem with our wives, they forbid us from having women conductors. You face so much during the day, the only thing you need is to be confronted with your angry wife about your conductor, a woman. Almost all drivers avoid this, and those who hire women, are hard hearted.
>
> (FGD, Dodoma)

Lack of clients/lack of earnings

When combining the two survey questions that pertain to daily earnings ("Few clients/competition/price fluctuation" and "Payment delays/non-payment/some days we get nothing"), a number of respondents cite financial concerns as an important challenge. The issue of financial survival permeates and is interwoven throughout the testimony in FGDs.

> Sometimes you see that it's getting late and that you have not gathered enough money to submit to the Boss and for your family, therefore, you do "whatever" it takes to obtain enough money for the day, there is no way out.
>
> (FGD, Daladala Dodoma)

Social protection services facilitated by transportation associations

This section discusses social protection as provided by the informal transportation sector associations. It borrows from the Devereux and Sabates-Wheeler (2004) framework, which conceptualizes different social protection measures as being "economic" (protective, preventive, promotive) and "social" (transformative). Protective measures, that is, cash transfers, are not covered in this study, as this is generally not something that informal associations engage with and are deemed costly. Many policy-makers promote preventive measures (i.e. health insurance), promotive measures (access to micro-finance, vocational training), and transformative (legal, institutional, social, and cultural changes that counter exploitation, discrimination, and abuse) social protection measures.

As seen in Table 4.2, the majority of workers mentioned work-related benefits, such as safety measures and higher/more stable income (promotive measures), as being the most important benefit. Bodaboda particularly appreciated safety and security because they are more prone to theft of motorcycle and their parts, kidnappings, and abductions. Daladala cited higher income as a major benefit. The second-ranked benefit by both groups is how associations provide emergency financial cushioning for education, social, and medical needs (preventive measures). The third benefit was obtaining loans (promotive measures), for example, association guarantee purchases of motorcycles on credit. Finally, voice and representation (transformative), as associations become consultative forum for members to dialogue and negotiate with authorities and other stakeholders.

Generally, both groups (daladala and bodaboda) mentioned the same association benefits, though with some differences between daladala conductors and drivers. Conductors usually depend on the goodwill from the drivers to hire and retain them, so membership in an association offers a chance to appeal to the association if a driver intends to dismiss them. Owners of buses, unless it is a family business, do not recognize conductors.

Conversely, associations are viewed as being ineffective in providing adequate support for health coverage and pensions and access to services such as water and toilets.

Preventive measures

Preventive measures aim to avert future situations that could lead to deprivation, for example, retired workers lacking sufficient savings and income, and a worker who falls ill, but cannot afford healthcare. In Tanzania, informal workers may contribute to the NSSF, NHIF, CHF, and private insurance companies such as Jubilee, AAR, etc.

As seen in Table 4.1, only 17 percent of bodaboda and daladala workers reported having formal social protection, out of which the majority were enrolled in iCHF, followed by NHIF, NSSF, and private insurance, which is more or less in line with the enrolment pattern by the general population. Most of the formal social protection schemes are financed through the contribution of employees and employers, a challenge for daladala and bodaboda operators. Most owners of vehicles do not consider daladala and bodaboda operators as their employees and therefore are not legally compelled to make mandatory contributions. Those who do have formal social protection either pay for it from their own resources or have willing employers. Out of those surveyed, 38 percent pay out-of-pocket for health services and medicines, again a reflection of the out-of-pocket payments made by the general population (Ministry of Health, 2015; World Bank,2020).

There are various reasons why informal transportation workers do not contribute to health insurances with minor variation across bodaboda and daladala operators as seen in Table 4.3. First, the vast majority (90 percent) have paid no attention to health insurance and how it works, due to several reasons, but mainly, they had never encountered serious health threats. Further reasons mentioned were enrolment procedures that were too complicated and too expensive.[13]

Under the NHIF Mutual Plan, also known as 'KIKOA', packages were tailored to "mutual groups" such as SACCOs, VICOBA, and bodaboda groups with 20 or more members. The more affordable annual premium of TZS 76,800 (USD 33.63) per person motivated several associations to try to register. However, daladala FGD informants recounted how difficult it was to apply under the 'KIKOA' program because NHIF considered them to be employees of the bus owners. They took their case to the higher authorities and were granted permission by the minister of health to enrol in the scheme. Due to adverse selection, the KIKOA program was terminated by NHIF towards the end of 2019. Instead, individual households can voluntarily enrol in NHIF for an average annual premium of approximately (USD 357–972), per family of six, which is beyond the means of informal transportation workers who, as seen in Table 4.1, earn an average of USD 10 per day.

The premium for iCHF is far more affordable than that of NHIF. A single premium for a household of six members is TZS 150,000 (USD 17.38) in Dar es Salaam, and elsewhere, in the country, it is TZS 30,000 (USD 13.13). Informants complained that the CHF does not always provide the services anticipated, often requires members to pay out of pocket for some expenses, and medicines are often not available. The limited services are also a problem as observed in FGD:

Table 4.3 Reasons for not contributing to health insurance

	Bodaboda	*Daladala*	*Total*
Never paid attention	90.0%	85%	89.6%
It is too expensive	1.1%	7.1%	1.2%
I prefer to save up myself	1.1%	0.0%	0.6%
Low quality of the insurance services	1.1%	0.0%	0.6%
Procedures too complicated	6.7%	9.8%	6.7%
No knowledge about insurance	0.0%	3.4%	1.2%
Total			100%
Association members	90	73	163

Source: Author's elaboration based on the project survey data.

> You can go with CHF card and think that they will assist you, but you are told that it does not cover certain services, and thus required to pay for such services, so what is the need of the card? Imagine time lost with long ques in hospital with the card. With cash, services move fast and you get time to go back to work....
>
> (Daladala FGD, Dar es Salaam)

Daladala and bodaboda groups and associations have arisen naturally, joining together by the social obligation to help people like themselves in need or under threat. Through these groups, money often is raised spontaneously to help a colleague in need. This is a unique cultural characteristic in Tanzania. The amount of informal fund-raising to support family, friends, and acquaintances in Tanzania is substantial. Many informal transportation worker groups put aside money in anticipation of serious illness or death of a member or close family member. This becomes an informal source of social wage and welfare protection for informal transportation workers, which helps manage the effects of precarious employment. The constitutions of many associations establish rules and regulations governing such savings.

The social dynamics among members make associations a far stronger means of SI than government sponsored programs, as observed in FGDs:

> —there are different types of accidents, many of them which are common, but in case of serious injuries, which requires exorbitant medical expenses, then our leaders will pass with the Daftari—Booklet—to other associations ... for contributions. Under such circumstances, you don't think twice, but to contribute ...We have solidarity as far as such issues are concerned...
>
> (FGD Bodaboda, DSM)

> The assistance we receive from the association is fast and immediate, which give us motivation to make contributions and pay fees as required because you know when it hits, you are covered. I tell you — being in this group gives me peace of mind as you know no matter what happened, you have someone behind your back to pick you up when you fall...
>
> (FGD, Daladala, Kijitonyama, Dar es Salaam)

Promotive measures

Promotive measures include training, strengthening human abilities, and household assets, which establish a basic foundation for households to persevere and grow. Recall from Table 4.1 that training on the job is more prominent in Dar es Salaam when compared to Dodoma because Dar es Salaam has a vast number of mainly bodaboda groups where one can start learning on the job as 'DAIWEKA' (read: day-worker). Informants during bodaboda FGDs often testified that they acquired their driving knowledge and skills by riding someone's motorcycle on parking lots. The lack of formal training on road signage, safe driving, and traffic laws and regulations may explain the high number of motorcycle accidents (AllAfrica, 2019; KII, Police, DSM).

The Vocational Education Training (VETA) and the National Institute of Transportation (NIT) are the main transportation training institutions. However, the training fees (tuition fees for the two weeks Driving Course is TZS 405,000 (USD 177.32) per participant) are relatively high; therefore, many drivers do not attend formal education courses. Even if the fees were lower, many informal transportation workers do not have adequate basic education to follow in the form of classroom-based theory, as noted by one FGD participant:

> You know it's difficult for us sometimes to go to such schools, because we are not educated, and then they start teaching you things that are not easy to understand, why don't they only teach us how to drive? Alright, they teach a little bit of driving towards the end, thus we pick up the real knowledge of riding from our colleagues.

Transaid (2015) reports theory-based driving school's curriculum in Tanzania. Workers also need financial literacy to manage personal finance and to monitor the financial affairs of their associations. There have been several cases of association leaders embezzling funds. One association member observed:

> ...Our association keep on changing over time, as we group then we dismantle the group, and in most cases is due to poor financial management issues. How come other associations have reached a point of buying their own Daladala, but our association leaders do not even call general meetings as are afraid people will ask about money and we will vote them out...
>
> (FDG, Daladala, DSM)

Associations have assisted in other promotive measures such as encouraging companies to establish lease and buy agreements with members. Also, as already highlighted, some associations were able to obtain loans to purchase buses. Motorcycle companies seek registered associations and offer members an opportunity to make a down payment and establish a schedule of weekly or monthly payments until the motorcycle is paid in full. Membership to an association provides motorcycle companies a degree of confidence that group support will ensure that payments are made as scheduled (FGD and KII,Daladala and Bodaboda, DSM and Dodoma). Some bodaboda and daladala associations have approached banks to create loan opportunities for their members (see CMPD, UWAMATEBU stories).

Transformative measures

Transformative measures aim to promote legal, institutional, social, and cultural changes that counter exploitation, discrimination, and abuse. In general, associations are involved in representation of their members as discussed in various sections in this chapter.

Informal transportation worker associations address abuse such as police harassment and lack of contracts from the employers, but their representational power is limited. Proactively, association leaders invite police to attend association meetings to build a positive relationship. They negotiate on behalf of individuals who have been harassed. However, leaders, who themselves are on the road earning a living, must handle the cases carefully not to expose themselves to retaliation. During FGDs, the efforts of associations were appreciated by members. Both daladala workers and bodaboda operators unanimously agreed that associations effectively represent them. Associations engage with police and authorities to settle members' problems, but generally provide limited services because they are too small and weak to influence authorities. The chairman of one of the larger daladala associations provided some insights as to why the contract issue is so difficult to resolve:

> ...there is no will on the side of the government. This is because most of the policy makers are owners of the vehicles and thus are not interested in listening to the grievances of the drivers. We have been talking for years, but it falls on deaf ears.

Some FGD participants expressed concerns that the push for contracts might backfire. If owners were required to pay salaries and benefits, this would upset the current arrangement that allows drivers to earn additional money after paying the set daily target, as noted by a daladala driver:

> I don't think that a Daladala owners can sincerely agree to pay salaries to drivers. If so, it'll pose a big problem for drivers. If you enter an agreement for a salary, the boss will raise the daily rate and impose stringent measures till you're left with nothing. If you don't comply, he can fire you by just saying – just park a vehicle. So, you can say this is not a formal job, unless they are forced to do so, and monitored for compliance, it won't work on our favour.

Discussion

The associations are important for workers in terms of provisions of social protection to members, albeit to a limited extent. This section reviews association entry conditions, makes comparisons between association members and non-members, and explores similarities between workers that are enrolled in formal SI versus those that are not.

Access to informal transportation sector associations

In order to be a member of an informal transportation worker association, one must have access to a vehicle[14] registered for business and a driving license. As shown in Table 4.2, 80 percent of respondents admit that there are barriers to enter the association. There may also be a probationary period, which daladala associations use to determine if a prospective member is a reliable business operator.

> ...when the driver wants to join our association, leaders welcome you, then they monitor you, for 3 or 6 months, if they see you as not troublesome, and that you can contribute up to TZS 4,000 (USD 1.75), for the total trips you make per day, they accept you and then they register you as member.
>
> (FGD Daladala, DSM)

The observation time also takes into account that a prospective member might not stay for long, "... you can register your bus to a route, but after four days, you may decide to change the route" (FGD, Daladala DSM).

Bodabodas are associated with reckless driving, crime, accidents (James, 2013; Xinhua, 2018) operating without licenses, but, in general, bodaboda associations use the initial period to observe behaviour.

> ...we started our association with a group of 18 members, but we found that other members are misbehaving, which was tainting our group. Then we changed our constitution and imposed tougher measures. We required IDs, set dress code, orderly parking of motorcycle to avoid competition for passengers. Such measures forced 10 members to drop out of the association; we are currently 8 only. Our business is booming, as we have created a good reputation in the community. We have consistent clientele, i.e. school children, elderly and grocery shopping for clients etc.
>
> (FGD, Bodaboda, Dodoma)

Most associations (68 percent) require payment of entry fees, which varies depending on the size of the association. Entry fees can be a deterrent to enter certain associations, especially for daladala drivers. One FGD informant observed, "...to join UWAMASTEBU you should have at least Tsh 250,000 (USD109.46), an exorbitant amount for many people to afford" (FGD, Daladala).

Other associations charge an entry admission fee of TZS 50,000 or more, while some association deter charging fees until a member has started earning money. Each driver pays to the association TZS 500–1,000 (USD 0.22–0.45) per trip. The monthly membership contribution varies between TZS 2,000 and 5,000 (USD 0.90–2.19).

Comparison between association members versus non-members

According to Pieterse (2008), collective action provides avenues for articulating group interests and agendas and for claiming rights. Some associations offer members access to group health care benefits and other perks that non-members lack. Table 4.4 shows the comparison between members of associations and non-members along a number of characteristics and other factors based on the survey data. As recalled from Table 4.2, SI enrolment and health insurance

Table 4.4 Differences in key workers characteristics by association member status

	Member	*Not Member*	*Difference*	*t-Value*
Social insurance enrolment	0.18	0.14	0.05	0.93
Health insurance coverage	0.17	0.14	0.03	0.57
Dar	0.41	0.40	0.01	0.13
Dodoma	0.59	0.60	−0.01	−0.13
Gender (male = 1)	0.98	1.00	−0.02	-1.27
Married	0.54	0.48	0.06	0.86
Local	0.48	0.55	−0.07	-1.01
Age	32.61	31.45	1.17	1.08
Mean daily earnings (current USD)	10.25	9.12	1.13	0.52
Assets (house and/or land)	0.42	0.34	0.07	1.11
Primary incomplete	0.04	0.06	−0.02	-0.76
Primary complete	0.67	0.63	0.04	0.58
Secondary or above	0.29	0.31	−0.02	−0.26
Training course	0.26	0.23	0.03	0.48
Training on job	0.15	0.16	−0.01	-0.16
Self-taught	0.59	0.61	−0.02	−0.31
Wage-worker	0.33	0.30	0.03	0.42
Own-account	0.55	0.69	−0.14	−2.12**
Micro-business	0.12	0.01	0.11	3.06***
Observations	250			

Notes: <0.01***, <0.05**, <0.1*.
Source: Author's elaboration based on the project survey data.

coverage were only slightly higher for members than for the full sample, indicating that membership is only marginally conducive to enhanced insurance, which is confirmed in Table 4.4 indicating no significant difference along that dimension.

Across most dimensions, association members do not differ significantly from non-members. They have similar educational profiles, the majority are Primary School leavers, but FGD discussions indicated that more educated workers are joining associations with the intentions of taking over leadership roles (FGD, Dodoma, and Dar es Salaam), in order to raise their visibility in leadership, and eventually run for political offices.

In terms of worker types, wage-workers are equally likely to be association members as well as non-members, but own-account workers are more likely not to be association members. This is because they often have their own means to obtain the kind of support that associations provide. Oppositely, micro-businesses (47.6 percent daladala, 4.8 percent conductors; 47.6 percent boda-boda) might join associations because they provide a new business owner with contact to other local businesses that facilitates integrating into the community and access to financial services, that is, SACCOs. Schleifer and Nakagaki (2018) report that, as entrepreneurs grow, they seek larger, commercial bank loans. One of the requirements for micro-businesses to access loans from the financial institutions in Tanzania is to belong to registered associations.

Despite the statistical insignificance, members of associations tend to earn slightly more than non-members. Members noted that associations help them be more financially disciplined. One FGD informant commented, "The money discipline put within our associations enable us to earn, and utilize fund properly. At the end of the day, you have more money than the guys who owns their motorcycles as they lax because have no submissions" (FGD Bodaboda). This was further illustrated by an FGD participant:

> ...associations teach financial prudence, but people belittle us, even my own brother refers my work in a derogatory manner. [wewe Bodaboda tu], and yet I earn more than he does, despite working for the government. Ironically, he borrows money from me.
>
> (FDG, Bodaboda, Dar es Salaam)

Comparison between formal social insurance of association members versus non-members

Table 4.5 compares respondents in the informal transportation sector who are enrolled in formal SI (including health insurance and pension) with those who are not enrolled. Again, whilst association members are more likely to have social insurance (SI) than not, the difference is not significant. Workers in Dar are significantly more likely *not* to have formal insurance, whereas the opposite is the case in Dodoma. One reason is that SI schemes are more expensive in Dar es Salaam than in Dodoma. In Dar es Salaam, iCHF costs TZS 150,000 (USD 65.67) per household of six/year; the same size household pays TZS 30,000 (USD 13.13) in Dodoma and other areas of the country. The reason is that general earnings and costs are higher in Dar es Salaam because of high population density (1,786 per km^2) (NBS, 2014b; NBS, 2016), which poses health risk, raising higher health seeking behaviour. Furthermore, iCHF objective includes access and coverage of the poor and vulnerable.

The results also show that the completion of Secondary school or above increases the likelihood of enrolment, while those who only attended primary school lag behind in enrolment. This supports the observation by Zajacova and Lawrence (2018) that more education increases the ability to comprehend the preventive value of health insurance.

Finally, own-account owners are more likely *not* to have formal social insurance because they have resources to provide alternative. Also, poor membership to associations excludes own-account workers from access to affordable health schemes tailored to associations members, for example, KIKOA, whilst wage-workers who have a comparatively higher associations membership share are able to benefit more from group health schemes.

To improve overall enrolment rates, formal SI packages must include appropriate contribution methods, acceptability measures, and benefit packages that befit the needs of informal transport sector workers. This ought to consider

Table 4.5 Differences in key workers characteristics by social insurance enrolment

	Social Insurance	*No Social Insurance*	*Difference*	*t-Value*
Association member	0.71	0.64	0.07	0.93
Sacco/vicoba/chama	0.19	0.13	0.06	0.94
Work-related association	0.52	0.50	0.03	0.34
Women/youth/religious	0.00	0.01	-0.01	-0.64
Work-related	0.33	0.26	0.07	0.91
Loans	0.24	0.15	0.09	1.42
Voice and representation	0.05	0.04	0.00	0.12
Social cushioning	0.10	0.18	-0.09	-1.38
Dar	0.26	0.44	-0.18	-2.12**
Dodoma	0.74	0.56	0.18	2.12**
Gender (male = 1)	0.98	0.99	-0.01	-0.77
Married	0.57	0.51	0.06	0.73
Local	0.55	0.50	0.05	0.56
Age	33.98	31.85	2.13	1.54
Mean daily earnings (current USD)	10.91	9.65	1.27	0.46
Assets (house and/or land)	0.43	0.38	0.04	0.53
Primary incomplete	0.00	0.05	-0.05	-1.53
Primary complete	0.52	0.68	-0.16	-1.98**
Secondary or above	0.48	0.26	0.21	2.76***
Training course	0.33	0.23	0.10	1.40
Training on job	0.12	0.16	-0.04	-0.72
Self-taught	0.55	0.61	-0.06	-0.70
Wage-worker	0.48	0.28	0.19	2.47**
Own-account	0.40	0.64	-0.23	-2.87***
Micro-business	0.12	0.08	0.04	0.90
Observations	250			

Source: Author's elaboration based on the project survey data.
Notes: <0.01***, <0.05**, <0.1*.

the fragility of the associations and thus the need for capacity-building of such associations.

Conclusion

This chapter assessed the relevance of informal transport associations for the provision of social protection to daladala and bodaboda workers. Despite poor regulatory environment, these workers are required by law to comply with formal regulatory requirements, such as obtaining drivers' licenses, paying vehicle insurance, parking in designated areas, wearing helmets, and plying specified routes. The vehicle owners require the drivers of daladala and bodaboda to pay a specified daily amount. Vehicle owners provide no social protection and do not comply with labour laws, and the government does not consider the informal transportation workers as parties under employer–labour relations regulations. Therefore, informal transportation workers view themselves as operating a business. Generally, informal transportation workers in Tanzania operate in both a world of formal and informality.

Informal transportation worker associations have emerged as a means of self-regulation in the face of government inability to regulate the sector, and members realize social protection benefits that derive from organizing themselves in associations. Power resource analysis shows that informal transport associations in Tanzania have inadequate structural power due to an oversupply of workers that gives vehicle owners the ability to set the terms of employment. The inability and unwillingness of the government to intervene strengthen the negotiation position of the owners of vehicles.

The workers demonstrate associational power by voluntarily forming informal sector associations. These associations provide preventive and promotive services such as provision of partial financial support to members who fall ill or victim to accidents, educating members about formal health insurance and pension schemes, spearheading driver's trainings that facilitate getting licenses for those who don't have them, assisting access to loans from banks, and providing short-term unemployment opportunities. Despite shortfall, informal transportation workers enjoy strong support from customers along their routes, and the workers on occasion have used this societal power to gain the attention of political leaders. But informal transport worker association lacks adequate institutional power to force government agencies to provide relevant social welfare programmes and schemes.

Generally, associations have a better comprehension of the felt needs, priorities, and contributing capabilities of the informal transport workers when compared with the formal sector institutions. The formal SI schemes offered by the government, that is, KIKOA, have not been consistent and successful in their outreach to the informal sector. The NHIF offers packages that are costly, and the CHF packages are deemed ineffective, as they fail to meet the needs of workers. The capacity of informal transportation associations must be enhanced in order to provide them the strength to be included in the policy-making process.

Notes

1 More than 70 percent of Dar es Salaam residents reside in unplanned areas with no convenient access to formal transportation and therefore rely on the informal transportation system (Ka'bange et al., 2014).
2 The terminated funds were the Local Authority Pension Fund (LAPF), Public Services Pension Fund (PSPF), Government Employees Pension Fund (GEPF), Public Pension Fund (PPF), and the National Social Security Fund (NSSF).
3 Poor coverage. According to Jacob and Pedersen (2018), only 1 percent of the population is covered by the NSSF.
4 Teachers, Hotels, Doctors, etc. left COTWU to form their sector-specific unions.
5 Interviews with COTWU-T IEDO, Training Manager, and Secretary General.
6 Source: Interviews with the association Chairman, and FGD Discussions, DSM.
7 UWAMAWIKI is composed of three routes associations UWAMASTEBU, UWAMAMAPO, and UWAMATEBU.
8 While they are not included in formal structures of the District, a representative from their group is involved in meetings and workshops pertaining to issues relevant to them, including land planning and taxation.
9 For daladala operators, LATRA sets the rates.

10 High level of self taught, contributes to poor adherence of rules and regulations of road safety thus contributing to higher rates of road traffic crashes.
11 Requires each LGA to set aside 4 percent of its revenue collection as a no-interest loan to youth.
12 Referring to the first mobile cellular network in Tanzania.
13 The same sentiments were echoed during FGDs in both DSM and Dodoma.
14 Vehicle refers both to motorcycles and daladala buses.

References

AllAfrica (2 September 2019) Tanzania: Road accidents down 27% – Report. *AllAfrica*. Available from: https://allafrica.com/stories/201909020347.html.

Cervero, R. (2000) *Informal transport in the developing world*. United Nations Center for Human Settlements, UN-Habitat. Available from: http://mirror.unhabitat.org/pmss/getElectronicVersion.aspx?nr=1534&alt=1.

Devereux, S. & Sabates-Wheeler, R. (2004) *Transformative social protection*. *IDS Working Paper 232*. Brighton, Institute of Development Studies.

Equity Bank loans assist members purchase their own motorbikes (conversation with CMPD Deputy Secretary, January 12, 2019).

Goodfellow, T. & Titeca, K. (2012) Presidential intervention and the changing 'politics of survival' in Kampala's informal economy. *Cities*. 29 (4), 264–270. Available from: doi:10.1016/j.cities.2012.02.004.

ILFS (2014) *Integrated labour force survey in Tanzania*. Dar es Salaam, National Bureau of Statistics.

ILO (2004) *Tanzania employment and labour relations Act, No. 6 of 4 June 2004*. International Labour Organization. Geneva, International Labour Office. Available from: https://www.ilo.org/dyn/natlex/docs/ELECTRONIC/68319/104204/F-894240970/TZA68319.pdf.

ILO (2018) *Social protection for older persons: Policy trends and statistics 2017–19*. International Labour Organization. Geneva, International Labour Office, Social Protection Department. Available from: https://www.ilo.org/secsoc/information-resources/publications-and-tools/policy-papers/WCMS_645692/lang--en/index.htm.

Jacob, T. & Hundsbæk Petersen, R. (2018). *Social protection in an electorally competitive environment (1): The politics of Productive Social Safety Nets (PSSN) in Tanzania*. ESID Working Paper 109. Manchester: Effective States and Inclusive Development Research Centre, The University of Manchester.

James, B. (17 August 2013) Inside boda boda crime syndicates. *The Citizen*. Available from: http://www.thecitizen.co.tz/News/national/Inside-bodaboda-crime-syndicates/1840392-1958102-fqea1k/index.html.

James, B. (2 November 2014) Terror as boda boda theft rises. *The Citizen*. Available from: http://www.thecitizen.co.tz/news/national/terror-as-bodaboda-theft-rises/1840392-2507796-54duqh/index.html.

Ka'bange, A., Mfinanga, D. & Hema, E. (2014) Paradoxes of establishing mass rapid transit systems in African cities. A case of Dar es Salaam Rapid Transit (DART) system, Tanzania. *Research in Transportation Economics*, 48. doi:10.1016/j.retrec.2014.09.040

Mbegu, S. & Mjema, J. (2019) Poverty cycle with motorcycle taxis (boda-boda). Business in developing countries: Evidence from Mbeya–Tanzania. *Open Access Library Journal*. 6 (e5617). Available from: doi:10.4236/oalib.1105617 [Accessed 2 March 2020].

Membership ID Cards protects informal transport workers from harassment (FGD with MMK, February 22, 2019)

Ministry of Health (17 March 2015) *Tanzania health financing strategy 2015–2025: Path towards universal health coverage, 3rd draft.* Ministry of Health and Social Services, Dar es Salaam, unpublished.

NBS (2014a) *Mainland Tanzania: Formal sector employment and earnings survey – analytical report 2013.* National Bureau of Statistics of Tanzania. Dar es Salaam, Ministry of Finance.

NBS (2014b) *Population and housing census: Basic demographic and socio-economic profile, Dar es Salaam region.* National Bureau of Statistics of Tanzania. Dar es Salaam, Ministry of Finance.

NBS (2016) *The 2012 population and housing census.* National Bureau of Statistics of Tanzania. Dar es Salaam, Ministry of Finance.

PDL (2014) *Labour and employment law overview for Mainland Tanzania (unpublished).* People Dynamics Limited (rev. September 2014).

Pieterse, E. (2008) *City futures: Confronting the crisis of urban development.* London & New York, Zed Books.

Rizzo, M. (2011) 'Life is war': Informal transport workers and neoliberalism in Tanzania 1998–2009. *Development and Change.* 42 (5), 1179–1206. Available from: doi:10.1111/j.1467-7660.2011.01726.x.

Rizzo, M. & Atzeni, M. (2020) Workers' power in resisting precarity: Comparing transport workers in Buenos Aires and Dar es Salaam. *Work, Employment and Society.* 34 (6), 1114–1130. Available from: doi:10.1177/0950017020928248.

Schleifer, M. & Nakagaki, M. (2018) *Empowering women entrepreneurs in Bangladesh.* In: Strategies for policy reform volume 3: Case studies in achieving democracy that delivers through better governance. Washington D.C., Center for International Private Enterprise.

Schmalz, S., Ludwig, C. & Webster, E. (2018) The power resources approach: Developments and challenges. *Global Labour Journal.* 9 (2), 113–134.

SID (2012) *Informal transport sector workers in the GHEA: A perspective from the driver's seat.* The SID Forum, Society for International Development. Available from: http://www.sidint.net/docs/RF22_Informal_Transportation.pdf.

Transaid (2015) *Tanzania motorcycle taxi rider training: Assessment and development of appropriate training curriculum.* Transaid. Thame, UK, Cardno Emerging Market (UK), Ltd. Available from: http://www.research4cap.org/Library/TRANSAID-2015-MotorcycleTaxi-Training-FinRept-AFCAP-TAN2015E-AnnexA-Curriculum-v150616.pdf.

TUCTA (2015) *Draft policy statement for the informal sector.* Secretary General, Msigwa, Trade Union Congress of Tanzania (TUCTA), Dar es Salaam.

UNICEF (2018) *Health budget brief.* UNICEF – Tanzania Mainland. Available from: https://www.unicef.org/esa/media/2331/file/UNICEF-Tanzania-Mainland-2018-Health-Budget-Brief-revised.pdf.

URT (2003) *National transport policy.* Ministry of Communication and Transport, Tanzania. Available from: https://www.tanzania.go.tz/egov_uploads/documents/National_Transport_Policy_2003_sw.pdf.

URT (2004) *The employment and labour relations act of 2004.* United Republic of Tanzania. Available from: https://www.tanzania.go.tz/egov_uploads/documents/Employment%20and%20LAbour%20Relation%20Act_1.pdf.

Vanguard (7 November 2018) How motorcycle accidents killed 800 in Tanzania. *Vanguard*. Available from: https://www.vanguardngr.com/2018/11/how-motorcycle-accidents-killed-800-in-tanzania.

Wang, H., Juma, M.A., Rosemberg, N. & Ulisubisya, M.M. (2018) Progressive pathway to universal health coverage in Tanzania: A call for preferential resource allocation targeting the poor. *Health Systems & Reform*. 4 (4), 279–283.

World Bank (2020) *Out-of-pocket expenditure (% of current health expenditure) – Tanzania*. World Bank. Available from: https://data.worldbank.org/indicator/SH.XPD.OOPC.CH.ZS?locations=TZ.

Xinhua (2018) Motorcycle accidents in Tanzania kill 800 people annually. *Xinhua*. Available from: http://www.xinhuanet.com/english/africa/2018-11/07/c_137589980.htm.

Zajacova A. & Lawrence, E.M. (2018) The relationship between education and health: Reducing disparities through a contextual approach. *Annual Review of Public Health*. 39, 273–289.

5 Informal transport worker organizations and social protection provision in Kenya

Anne W. Kamau

Introduction

Kenya's transport sector is dominated by road transport which accounts for over 80 percent of total passenger traffic and 76 percent of freight. The informal road transport modes include public service modes such as the paratransits commonly known as 'matatu' and the motorcyclists commonly known as bodaboda. The two are common modes of passenger transport in Kenya (Behrens et al., 2017; Mutongi, 2017). Paratransits or matatus are also referred to as public service vehicles (PSVs) and are often used interchangeably, even though the term PSV includes other modes like passenger transport motorcycles. This chapter focuses on paratransit and motorcycle sub-sectors workers in the urban areas of Nairobi and Kisumu cities in Kenya. The two modes employ a large pool of informal workers who directly and indirectly depend on the sector. The matatus sub-sector direct beneficiaries include the vehicle owners, drivers, conductors, fleet managers, stage managers, callers, and loaders, while the indirect workers include mechanics, painters, graffiti artists, and cleaners (Wright, 2018). For the motorcycles sub-sector, the key actors are the riders and owners. Despite many players in the transport sector, there is no reliable data on number of workers. Spooner and Whelligan (2017) estimated that, in 2017, over 100,000 PSVs had been registered, while Opondo and Kiprop (2018) estimated that about 1,393,390 motorcycles were registered in 2018.

This chapter focuses on informal transport workers in the paratransits and motorcycles sub-sectors in Kenya, their social protection coverage, and the role of associations. The chapter is based on data obtained through research undertaken in Nairobi and Kisumu between June and December 2018. The collaborative study used quantitative and qualitative methods in data collection that included a survey of 200 informal public transport workers, focus group discussions with workers and vehicle owners, and key informant interviews with representatives of government organizations, workers' associations, and unions. The chapter starts with a reflection on governance of the sector, followed by a discussion on formality and informality. It then reflects on workers' associations and an analysis of the study findings on workers characteristics and associations' membership. The chapter uses the power resources analytical framework to discuss the role of

DOI: 10.4324/9781003173694-5

associations in promoting workers access to social protection and concludes with reflections on the two sub-sectors.

Transport sector governance

The public transport sector in Kenya is privately organized and governed through a top-down structure. The sector is regulated through complex multi-layered and multi-agency systems that have formal and informal regulations and controls. Different agencies regulate different aspects of the sector, albeit sometimes with overlapping and conflicting roles (World Bank, 2019). The most prominent agency is the National Transport and Safety Authority (NTSA). Its mandate is to develop and implement traffic regulations, registration, and issuance of motor vehicle licenses, inspection and certification, and advising government on national road transport sector policy issues. NTSA also requires PSVs to be installed with digital speed governors, seat belts to enhance passenger safety, fleet monitoring system for vehicles travelling at night, and to undergo annual inspection (ROK, 2012a, 2012b). Enforcement of these regulations is undertaken jointly by the NTSA, the National Police Service (NPS), and the motor-vehicles inspection units. Other government agencies that play crucial roles in the sector operations include the State Department for Co-operatives of the Ministry of Industry, Trade and Co-operative, and the Criminal Investigation Department (CID).

An example of multi-agency sector regulation is the implementation of the NTSA Legal Notice 161 of 2012 that requires all PSVs to join transport Savings and Credit Co-operative Societies (SACCOs) or transport management companies (TMCs). The SACCOs and TMCs are registered by the State Department for Co-operatives and are expected to manage vehicles registered under them and clear workers for licensing by NTSA and issuance of Certificate of Good Conduct by the CID (NTSA, 2014). This chapter uses the term SACCOs to refer to both the SACCOs and TCMs as there is no clear conceptual difference between the two terms. Initially, most operators had registered companies which were later transformed into SACCOs in order to comply with the NTSA requirements. There is no reliable data on the number of transport SACCOs and companies registered in Kenya. Estimates suggest that Nairobi alone has about 692 registered SACCOs, but only about 272 are registered with the NTSA to operate (Omulo, 2020; Mwanza, 2021), suggesting that about 420 SACCOs that are in government records as registered are not licensed by the NTSA to operate.

Formality versus informality in the transport sector

The question of formality and informality generates varying views. The 'formal' perspective asserts that the sector has formal structures and regulations that govern its operations. The formalization process, according to this view, begins when operators and workers engage with the formal systems involved in inspection, licensing and certification of vehicles, and licensing of operating routes and workers. An example is the formal registration process that requires matatu operators to join SACCOs to be licensed by NTSA and the workers to get clearance from the CID so as to be licensed to work in the sector. Informality starts at the level of operations which are mainly informal, according to this view. As argued, operators enter the sector through formal systems but become informal once they begin to work. The view that considers the sector as 'informal' bases its arguments on the precarious working conditions that characterize the sector such as job insecurity, lack of formal contracts, irregular and unreliable incomes based on targets, lack of leave days or official retirement, limited or no access to credit, and generally, weak representation. This is despite existence of legislation that would enhance formalization of the sector, for instance, the requirement to provide formal work contracts to drivers and conductors in the matatu subsector. There is thus a contradiction between the existing legal frameworks and the practice. Hence, the challenge in Kenya is not the absence of laws to protect workers but lack of enforcement and supporting systems, and this makes the sector prone to remain being informal.

Services offered by transport associations

Kenya's transport sector is largely association-based, and for the matatu's sub-sector, it is mandatory for PSV operators to join transport SACCOs. This is, however, not required of the motorcycle (bodaboda) operators. Nonetheless, they also belong to sub-sector-based associations. The informal transport workers' SACCOs and associations can play a role in ensuring workers' access to formal social protection. However, evidence suggests that most of these workers, like other informal sector workers, lack access to formal social protection, for instance, health insurance. Even when enrolled in social security schemes, they also depend on informal networks to meet their welfare needs. This chapter explores the role of transport associations in enhancing workers access to formal and informal social protection in line with the preventive, promotive, and transformative framework elaborated by Devereux and Sabates-Wheeler (2004).

Preventive

Preventive measures seek to avert deprivation or destitution, with a focus on risk mitigation and diversification, for instance, by providing or promoting access to social insurance. The transport sector is a high-risk sector that is prone to accidents, and workers are exposed to continuous risks. They face health and social challenges that include exposure to noise and air pollution, eye problems, fatigue, and exhaustion due to working for long hours and sometimes working in a hostile environment. To avert possible occupational hazards and the possible poor health, it is important for the sector's actors to invest in preventive measures. This investment could include enrolment of workers in social insurance schemes and improvement of the working environment. Examples of preventive measures include the NTSA Act of 2014 which requires SACCOs to employ workers on formal work contracts, provide them with leave days, and make statutory contributions to health and pension schemes. Employers are also expected to ensure health and safety of the workplace and comply with the Work Injuries Benefits Act (WIBA) No. 13 of 2011 (ILO, 2020). Also, the Kenya subsidiary legislation of 2020 requires drivers to undergo medical examination before being issued with driving licences to assess their fitness (ROK, 2020).

The National Health Insurance Fund (NHIF) and the National Social Security Fund (NSSF) are examples of social security programmes that transport workers can enrol in. Survey findings in this study showed that 47 percent of workers had formal insurance (slightly higher for health insurance coverage), with the vast majority referring to the NHIF, followed by the NSSF and other types including the one-year universal health pilot programme rolled out in 2018. Discussions with key informants revealed that some workers enrol in insurance programmes, especially the NHIF, voluntarily without influence from the SACCOs. In some cases, though not common, SACCOs like the Kangemi Matatu Owners (KMO) in Nairobi, Kihomi Sacco in Kisumu and Kisumu Ahero Mowouk Transport Company encourage workers to enrol in health insurance programmes. Further, some SACCOs make daily deductions from the workers' wages and remit the pooled savings to the schemes on monthly basis on behalf of workers. In this case, the employer enables the workers to pay for social security coverage but do not pay for the workers' coverage. Nonetheless, most workers are reluctant to join employer-linked social security programmes due to fear that their job mobility might be restricted. Low awareness about social protection schemes and superstition that taking health insurance is a bad omen leaves many workers without coverage. Results also indicate that most workers belong to associations that cater for their welfare needs like bereavement, sickness, and school fees support. Such support is often limited and may not cover full costs, for instance, in case of hospitalization as would be the case with NHIF.

Promotive

Promotive measurers are aimed at enhancing real incomes and the productive capacity of workers. In this study, such measures include training and capacity-building of workers, and those that enhance workers' financial investments and savings. Examples of promotive measures were evident in the study, for instance, where SACCOs facilitated workers training on financial literacy, road safety, and investment in social protection. Some support workers' access to credit from financial institutions like the Kenya Union of Savings and Credit Cooperatives (KUSCCO), and banks like Equity and Family banks. There are also those that promote workers' investment through savings, share-holding, or contributions. The money is used to purchase SACCO or company vehicles, individual, or association motorcycles or to access loans. Most associations and some SACCOs pay annual dividends to members from these investments. In most cases, even with informal associations, the annual sharing of dividends is common, and members are free to use their money in the way they prefer, like paying school fees and meeting festivities cost. However, hardly is the money used to pay for social security. Hence, some key informants emphasized the need to sensitize workers about investing in their own social security. As noted by Kamau et al. (2019), informal sector workers have different peak seasons that social security schemes could target to expand coverage. In the case of the transport sector, the peak season is in December, and agencies like the NHIF and NSSF could engage workers during this period to encourage them to pay for social security.

Transformative measures

A transformative social protection approach addresses issues of social equity and inclusion by focusing on policies and legislations as well as taking measures that include collective action for workers' rights. Results showed weaknesses in the transformative role of associations, especially in enabling workers' access to social protection. In the survey, only 9 percent of association members reported representation and voice as a key benefit, and this was mainly about engagement with authorities to address challenges facing workers like harassment by authorities, petitioning the government to rescind decisions considered a threat to the sector such as removal of 14-seater matatus from the sector, or lobbying the government to undertake interventions to improve the sector operators such as improving roads. Some SACCOs also intervene to discourage the rent-seeking behaviour which involved payment of bribes to police and cartels, as noted by 61.4 percent of surveyed workers. Some SACCOs instruct workers not to pay bribes and instead to contact them if they have problems with authorities. On the extreme side like in Kisumu, some SACCOs encourage workers not to wear uniforms to avoid being easily identified by authorities. Despite this, the workers rarely contact their associations on policy, regulatory, or work-related issues affecting them. From the survey, 23.5 percent of workers had contacted their SACCOs to get welfare support, 6.5 percent for associations to negotiate with the authorities, and 4.5 percent to get help on work-related issues like better terms of employment and salary payment, increase, or promotion.

There is no common bargaining agreement on workers' rights, and labour issues are hardly pursued. Formal work contracts are hardly provided to workers despite being in the sector for long. In the survey, only 1.5 percent of workers had written contracts, 18.5 percent had verbal contracts, 23 percent day contracts, and the rest (23 percent) had no contracts at all. Hence, workers have no legal protection and can easily lose jobs if they fail to meet targets or disagree with their employers. Nonetheless, some SACCOs such as Kihomi in Kisumu and Kisumu Ahero Mowouk Transport Company (KAMTCO) employ workers on formal work contracts if they work for the vehicles that are franchised to the SACCOs. As observed:

> There is no employee protection from the employer in case of a dispute. The employee has no job security since there is no contract. They also do not enjoy benefits such as leave allowance. The income in this sector is unreliable.

> Among the boda-boda group the sector is informal and they depend on chamas in case of illness or accidents. There are those who have NHIF cards through individual arrangements.
>
> (FGD, Nairobi)

Workers' representation is weak, and most SACCOs and vehicle owners are reluctant to support workers to join unions. The common trend among the workers is to change or quit their jobs if they are dissatisfied as opposed to contacting their employers to voice their concerns. This results in high workers turn-over and the consequent low social security coverage. The exit option is a weaker strategy as it does not help to create stable work environments. McClean, Burris, and Detert (2013) noted that "voice and exit are directly inversely related" and that often employees have two distinct choices which entail (i) choosing voice or silence and (ii) choosing to stay or exit their organization. This, according to McClean et al., implies that "employees may sometimes speak up and remain in their organization irrespective of how much or how quickly things change", or they "may speak up and subsequently exit the organization because of what happens (or fails to) in response to voice". The fact that workers can be dispensed at any time by their employers makes them opt for the exit option. This resonates with McClean et al.'s view that voice does not automatically make things better or worse for those who speak up or anyone else, and as noted by FGD participants in Kisumu,

> Jobs in the sector are insecure and a driver considers himself as working by having the key. It is common for the owner of the vehicle to inform the driver that he is using it for other things like travel, only for the driver to find it on the road with another driver.
>
> (FGD, Kisumu)

Transport workers' characteristics

This study covered matatus workers – drivers and conductors – who comprised 58 percent of the survey sample and motorcycle operators (*bodaboda*) who made up 42 percent. Transport workers are generally young with a mean age of 36 years and most are married (86 percent). Around 29 percent were born in the city where they work, though this is much lower for Nairobi at 13 percent, indicating the high extent of urban migration which is most likely work related. As seen in Table 5.1, membership in associations is 67 percent and is substantially higher in Kisumu compared with Nairobi. By contrast, formal social insurance enrolment, which is 47 percent overall, is higher in Nairobi compared with Kisumu, and for health insurance coverage, the difference between the two cities is even larger, with Nairobi having a coverage rate of 59 percent.

The vast majority of transport workers are men, though with 7 percent women in Kisumu. In the survey, all drivers and motorcycle riders were men. The absence of women is linked to the nature of the transport sector work which

Table 5.1 Key worker characteristics

	All		*Nairobi*		*Kisumu*	
	Mean	*SD*	*Mean*	*SD*	*Mean*	*SD*
Association membership*	0.67	0.47	0.60	0.49	0.78	0.42
Social insurance enrolment	0.47	0.50	0.52	0.50	0.42	0.50
Health insurance coverage	0.50	0.50	0.59	0.49	0.37	0.49
Male	0.97	0.17	1.00	0.00	0.93	0.26
Age	36.16	8.89	35.75	9.52	36.73	7.97
Married	0.86	0.34	0.84	0.37	0.90	0.30
Born in local city/town	0.29	0.45	0.13	0.34	0.51	0.50
Mean wage in USD	10.95	7.09	12.42	7.98	8.93	5.00
Assets ownership (house, land)	0.28	0.45	0.27	0.44	0.29	0.45
Primary education incomplete	0.11	0.31	0.07	0.25	0.15	0.36
Primary education completed	0.47	0.50	0.42	0.50	0.55	0.50
Secondary education completed	0.42	0.49	0.51	0.50	0.30	0.46
Professional training course	0.45	0.50	0.46	0.50	0.43	0.50
Training on job	0.27	0.45	0.28	0.45	0.25	0.44
Self-taught	0.28	0.45	0.26	0.44	0.32	0.47
Wage-worker	0.66	0.47	0.61	0.49	0.73	0.45
Own-account	0.28	0.45	0.31	0.46	0.25	0.44
Micro-business (with 1–2 employees)	0.06	0.23	0.08	0.27	0.02	0.15
Observations	200		116		84	

Source: Author's elaboration based on the project survey data.
Notes: Summary statistics based on the full sample. * Association member figures are from the random sample consisting of 152 workers. ** The median wage is USD 9.6 overall, USD 9.9 for Nairobi, and USD 7.4 for Kisumu.

involves a lot of shoving and jostling among operators and riders and is characterized by negative public perception and unfavourable working conditions (Wright, 2018). Other reasons for low participation of women in the sector include harassment and intimidation as well as historical and cultural reasons that discourage women from entering such 'risky' jobs (Kamau & Mitullah, forthcoming). This also mirrors the global trend where men dominate the sector (Wright, 2018). Nonetheless, women are increasingly taking up public transport jobs mostly as conductors.

The average daily income is about USD 11 and is around 25 percent higher in Nairobi compared with Kisumu. As expected, the median wage is lower than the mean (at USD 9.6 overall) due to the right skew of the distribution with a few heavy earners at the top pulling up the mean. For this reason, the median is often a more accurate measure of the general wage level. As for assets, only 28 percent own a house or land. In terms of education, 42 percent have attained secondary education or above (substantially higher in Nairobi), and 45 percent have attended a professional training course, whilst 27 percent have received training on the job. Most transport workers are wage-workers (66 percent), with the incidence being especially high in Kisumu at 73 percent. By contrast, the incidence of own-account workers is slightly higher in Nairobi (31 percent) than in Kisumu (25 percent). Finally, micro-businesses make up only 6 percent of the sampled workers. In the survey, the own-account operators were mostly motorcycle riders while the matatu operators had a higher representation among wageworkers. The relatively high rate of motorcycle ownership is attributed to low purchasing capital required for motorcycles ranging between KES 65,000 and 80,000 (Mwobobia, 2011). In contrast, owning a PSV requires high capital investment on top of which comes the cost of vehicle importation, registration, insurance, and general maintenance.

Associational landscape

Transport workers belong to different sub-sector-specific associations, for instance, matatu SACCOs or motorcycle associations, and even smaller informal welfare groups known as chamas *(Kamau et al. 2018)*. There are also umbrella associations that represent the SACCOs, owners, and workers. Examples include the Matatu Owners Association (MOA)[1] in which own-account operators belong to Matatu Workers Association (MWA), Matatu Crew Workers Association (MaCREW), and the Boda-boda Association of Kenya (BAK). Together, these SACCOs and the associations account for the vast majority of members who belong to associations (80 percent), followed by the worker, trade, and business-specific associations and unions (17 percent), and others like women and youth organizations and unions (3 percent). Interestingly, whilst the incidence of worker, trade, and business associations is higher in Kisumu, that of MOA, SACCOS, and *chamas* is substantially higher in Nairobi, thus indicating a difference in the type of and preference for associations across locations. There are also unions representing workers in the sector, and these include the Kenya Transport Workers' Union (TAWU), Matatu Workers Union (MWU), and the Public Transport Operators Union (PUTON). TAWU has representation in Nairobi and Kisumu, while MWU and PUTON are mainly in Nairobi. These unions are also allied to the International Transport Workers Federation (ITF).

As mentioned earlier, the survey data revealed that workers' membership in transport associations is relatively high at 67 percent, with 33 percent of workers not belonging to an association. Membership in transport associations in Kisumu is higher (78 percent) compared with 60 percent in Nairobi. The association enrolment-gap was associated with workers' reluctance to join SACCOs that could restrict their job mobility, failure by NTSA to involve operators in decisions relating to SACCOs formation and to set regulations that bind workers to remain for a specified period in SACCOs that facilitate their entry to jobs in the sector. As noted in the discussions, most workers perceive SACCOs merely as enablers for getting NTSA clearance and accessing jobs in the sector. Hence, as noted by a key informant, "PSV workers are not concerned about having social insurance and they comply with SACCO rules to get jobs".

Transport associations vary in membership size and structure and have formal and informal elements. The matatu associations, mainly the SACCOs, have a simple structure and are mainly geographical and route-based. The motorcyclist associations are largely multi-layered and hierarchy-based, with leadership structure being at the 'bases' or zonal levels, with further linkage to county and national levels. An example is the Boda-boda Association of Kenya (BAK) which is an umbrella association and has representation across the country including in Nairobi and Kisumu. In both Nairobi and Kisumu, bodaboda associations comprised 'base-groups' that had designated parking points. Each base has a leader and is represented at the zonal, county, and, in some cases, in the umbrella associations. For the motorcyclist, their associations are need-based, self-driven, and self-governed. Unlike matatu SACCOs, motorcycle grassroots associations create synergy among members. However, the PSV SACCOs are mostly formal and stable while motorcycle associations have greater flexibility but are shakier. Due to the high membership numbers, bodaboda associations are less stable and are occasionally marred by internal conflicts and splinter groups. They are also easily infiltrated by politicians who use them to get political mileage.

Table 5.2 presents selected summary statistics based on association members only. First, the table shows that association members have a slightly higher incidence of social insurance enrolment (49 percent) and health insurance coverage (52 percent) compared with the full sample of workers in Table 5.1 (47 percent and 50 percent, respectively). Again, both rates are significantly higher in Nairobi compared with Kisumu. In terms of association types, 80 percent of workers report being in SACCOs with close to 90 percent of workers in Nairobi. Regarding benefits, the majority of workers (46 percent) report loans as the main association benefit (especially in Nairobi) followed by work-related benefits (41 percent), voice and representation (9 percent), and finally, social cushioning (4 percent). Voice and representation are cited as more important among workers in

Table 5.2 Key worker characteristics, association members

	All		*Nairobi*		*Kisumu*	
	Mean	*SD*	*Mean*	*SD*	*Mean*	*SD*
Social insurance enrolment	0.49	0.50	0.57	0.50	0.38	0.49
Health insurance coverage	0.52	0.50	0.65	0.48	0.35	0.48
Association type						
Sacco	0.80	0.40	0.88	0.33	0.71	0.46
Work-related association	0.17	0.37	0.13	0.33	0.22	0.41
Other association	0.03	0.18	0.00	0.00	0.08	0.27
Benefit type						
Work-related	0.41	0.49	0.44	0.50	0.37	0.49
Social cushioning	0.04	0.20	0.03	0.16	0.06	0.24
Voice and representation	0.09	0.29	0.04	0.19	0.15	0.36
Loans	0.46	0.50	0.50	0.50	0.42	0.50
Barriers (yes = 1)	0.58	0.50	0.65	0.48	0.49	0.50
Association fee (yes = 1)	0.83	0.38	0.74	0.44	0.94	0.24
Observations	145		80		65	

Source: Author's elaboration based on the project survey data.

Kisumu (15 percent) which could be related to the activities of traders generally being more regulated in Kisumu, and hence more issues on which associations may intervene.

Table 5.2 also shows that most transport associations (58 percent) have entry conditions, particularly, in Nairobi. In most cases, membership in SACCOs is vehicle-based, and except in a few cases like Kisumu Ahero Mowouk Transport Company in Kisumu, workers are not direct beneficiaries. Instead, they are in the SACCOs merely to get jobs, since this is a requirement by the NTSA. The vast majority of associations (83 percent) require payment of a one-off entry fee, ranging between USD 500 and 1,000. For the motorcycle associations, the entry fee is in the hundreds. In addition, vehicle owners are required to pay daily contributions for SACCO management. The owners and workers also contribute daily savings that range from about 1.5 to 2 USD for PSVs and about 0.2 to 2 USD for motorcycles. This money is used to cater for welfare issues and loans provision to members and in some SACCOs to cover social security. Some SACCOs like the Lakebelt and BAK require owners and also encourage workers to make investments savings through share capital contributions that are pooled to buy assets like vehicles or motorcycles. In most cases, investment in shares is voluntary.

Highlights on selected transport associations

Matatu Welfare Association (MWA)

MWA was formed in 1996 and formally registered in 2000. It grew out of protest against a proposal by the minister for transport in 1996 to introduce installation of 'technographic black boxes' in PSVs in order to capture road crash data in the event of an accident. The workers protested the high cost of the gadget estimated as USD 700 (KES 70,000) per piece. MWA is SACCO-based, and its vision is to protect workers in the matatu industry. Membership entails payment of a one-off fee of USD 100 (KES 10,000) and a monthly fee of USD 50 (KES 5,000) per SACCO. Since formation, MWA has played a key role in transforming the industry through lobbying, advocacy, and engagement with government, policy-makers, and key stakeholders as well as addressing the challenges facing the industry.

The association has, over time, engaged with the government on issues related to organization of the sector and compliance with regulations and has developed initiatives to support workers. In 2004, for instance, with the introduction of new matatu rules, MWA organized a benchmarking trip for its members to travel to Uganda to learn about maintaining workers discipline, vehicle branding, and speed governor installation. MWA was also involved in the development of the Integrated Public Transport Policy. Despite this, discussions revealed that MWA has not done much to address labour issues affecting workers. The association does not provide or support members' access to formal social insurance nor labour protection as the individual SACCOs are expected to facilitate this.

Nonetheless, MWA supports members' welfare when need arises for instance in case of workers arrest or harassment by authorities on work-related issues (but not due to obstruction or overloading). The issue of supporting workers training and credit provision is also left to the SACCOs.

Kisumu Ahero Mowouk Transport Company, Kisumu

Kisumu Ahero Mowouk Transport Company (KAMTCO) is a limited company started in 2010 and is registered with the Ministry of Co-operatives. The company has 11 directors (all men). To become an official, one must have a vehicle in the company. At the time of the interview, the company had 114 PSVs that belonged to members and an additional seven that belonged to the company. The company has four women investors while the rest are men. Membership benefits include protection against harassment by authorities and negotiation with authorities in case of an accident or other problem. As noted by an official, "it is the face of the company that is seen not that of the individual investors".

In terms of fleet management, vehicle owners are required to sign franchising contracts with the company. However, not all vehicles are managed by the company. At the time of the interview, only 30 vehicles had been franchised to the company for management, while the rest were managed by the owners. The company had employed 89 workers who included drivers, conductors, and stage managers. About half of them (41), mainly from the franchised vehicles, had formal work contracts. Once employed, the drivers and conductors are put on a three months' probation. If confirmed, they get formal contracts and have a regular monthly salary of USD 300 and 200 (KES 30,000 and 21,000), respectively. Each vehicle is assigned three off-board workers (Callers) and is charged KES 450 maintenance fee daily of which 300 is used to pay the three Callers and 150 is for office management. In addition, each vehicle contributes KES 200 daily (about USD 2), from which KES 100 is for SACCO management and KES 100 for vehicle owner savings.

The company makes regular contributions to NHIF to cover enrolled workers, even though few are enrolled. Out of the 89 workers, only 13 drivers and 6 conductors had enrolled in NHIF.

The vehicle owners are responsible for paying for workers social security, but most are reluctant due to high staff turnover. The workers thus pay for their NHIF and NSSF contributions, but their enrolment is low due to fear among most workers of being tied to employer-linked schemes. The drivers and conductors also get paid maternity and paternity leave. The company has a burial and benevolent fund which covers the 89 employees who contribute KES 20 daily. In addition, the company has a welfare and investment SACCO whose membership is open to non-vehicle owners, including the workers so long as they are based in Kisumu. Not all the 89 members are members of the welfare SACCO even though data on the number of workers who belonged to the SACCO were not obtained. In the SACCO, members contribute a minimum of KES 2,000 monthly for shares. In some cases, the money is used to purchase a SACCO vehicle from whose profits the members get annual dividends. In addition, the SACCO caters for members' welfare issues like medical expenses of drivers and conductors who belong to the SACCO, bereavement, and children's education support through school fee loans. This means that workers who are on contract and are members of the SACCO can benefit both from the company and the SACCO.

Discussions with the company officials revealed that despite having a savings and investment SACCO, the workers and members have a poor saving culture. To address the challenge, the company organizes training for the members. For instance, it engaged the British-American Insurance Company (BRITAM) to educate members on importance of joining social insurance schemes, and the National Industrial Training Institute (NITA) to train members on financial literacy among other aspects. In addition, the company organizes exchange visits for workers so as to learn from successful SACCOs in Central Kenya like the 2NK.[2] Whereas the company is considered to be a success, it faces challenges like interference by politicians especially during political campaigns, worker harassment by authorities, and high staff turn-over. The refusal by some owners to franchise their vehicles to the company also weakens the company's management and control of workers.

Kaloleni Shauri Moyo Boda-boda SACCO

The SACCO was formed in September 2018 with about 120 members from seven bases. The number had grown to about 56 bases with over 1,000 members at the time of the interview. The idea to form the SACCO arose after some members visited Rwanda for a benchmarking trip sponsored by the County Government. Upon returning to Kenya, they formed the SACCO to implement the good practices that they had learnt, for instance, on riders' discipline, traffic regulations on passenger carrying capacity, having driving licenses, and keeping the work areas clean.

The SACCO has an open membership, and at the time of the interview, it also admitted members who did not own motorcycles. The SACCO was organized around zone groups (community-based) and below them were the 'bases' with the leadership structure having representation from different zones and bases. Each base is self-governing with own rules and regulations, that include determining the members' registration fee, although the minimum is KES 200. Some bases require members to pay up to KES 2,000 to join. Membership requirements include being over 18 years old and being registered with a base. At the time of the interview, the SACCO had about 40 women members, some of whom were not riders. Some owned motorcycles and had employed riders and some had businesses near the bases.

Registration is also required at the SACCO level. Here, members pay KES 200 registration fee, KES 100 daily to the SACCO, and KES 50 daily contribution. Once collected, the money is banked by the SACCO under its savings portfolio. Each group (base) contributes KES 750 per week of which KES 50 is for supporting office operations, KES 200 is put in members' savings, and KES 500 is for merry-go-round contributions that are used to give members loans. The officials estimated that, every month, the SACCO advances loans of up to KES 72,500 to members. From the onset, the SACCO aimed at empowering members to own motorcycles so as to stop being employed (*kodesha*). Within one year of formation, 67 members had received loans to purchase motorcycles. Additional financial support is sought from banks and cooperatives. For instance, the SACCO had acquired a loan from Equity Bank to purchase 10 motorcycles and from UNAITAS Savings and Credit Cooperative Society Limited to purchase 20 motorcycles.

The SACCO provides other benefits to members. For instance, supporting training on financial management, driving, and road safety as well as getting third-party motorcycle insurance through a local insurance company at a cost of KES 5,800, but the riders were not covered. Instead, they contributed KES 80 daily to cover for eventualities like accidents. At the time of interview, the SACCO had a welfare fund that catered for bereavement in case of death of a member or a family member, with each member contributing an additional KES 200 if a member was bereaved. In addition, each base was required to give matching contributions of KES 200. The funds covered funeral expenses like coffin, transport, and mortuary fees. The SACCO also had a school fees kitty which supported children of members who joined secondary schools.

A key challenge facing the SACCO as discussed with the officials is internal conflicts that sometimes lead to the formation of splinter groups. There is also interference by politicians, and this affects the SACCO management. As noted, officials and members who have close ties with political leaders approach them to get support to benefit themselves, and not necessarily, the workers. This undermines the SACCO leadership and causes conflict among the members.

The power resources framework

This chapter, like the rest of the book, is guided by the power resources approach (PRA) analytical framework which is based on the premise that, if organized, labour can successfully defend its interests by collectively mobilizing different power resources. Put differently, and as implied by Lund (2009), informal workers' organizations can be a gateway to the realization of informal workers' rights. The workers thus need not only to have rights but they also need to be aware about their rights and be able to claim, protect, and maintain them. These 'rights stages' can be achieved through use of different forms of power that include structural, associational, institutional, and societal power. In the transport sector, the workers' exercise of these powers varies by sub-sector. This chapter explores the different forms of PRA that are used by transport workers and examines whether belonging to associations enhances workers' ability to claim and defend their interests. For instance, the fact that matatu workers are mostly employed and belong to SACCOs, while the bodaboda workers are mainly own-account operators, means that different PRAs are used to protect their interests.

Structural power

As a form of labour power, structural power includes the marketplace, workplace, and regime-disruptive powers. The position of informal transport workers in the work- and marketplaces is weak mainly because the sector is under-valued, and workers are still to a large extent perceived as rogue workers. The large pool of available workers also means that workers are easily replaceable. However, the workplace bargaining power is stronger due to the large number of workers that increases their ability to disrupt operations. From the onset, workers enter into a sector that is not fully appreciated or recognized, and one that does not have systems that enable workers to easily access or claim their labour rights. The sector, being informal and without supportive structures, weakens the workers bargaining position. Most transport workers, especially the matatu workers and some bodaboda operators, are employed, and their work compensation and benefits are pegged on informal contracts. Except in a few cases, mainly for bodaboda operators who own the motorcycles, transport workers do not own the vehicles, and even though they control the income from the vehicles, their job security is not guaranteed. Equally, the SACCOs they belong to largely serve the interests of vehicle owners, even though, in a few cases, some SACCOs facilitate workers' access to social protection. The lack of work- and marketplace power means that workers' negotiation space is limited even for demanding for their right to social security. Thus, other than pushing for better working environment, most workers, especially in the matatu sub-sector, prefer to change jobs or employers, and this contributes to some workers' unwillingness to have employer-linked social security provision. In the discussions in Kisumu, some matatu workers indicated that they avoided enrolling in schemes initiated by their employers due to fear that this could curtail their job mobility. Hence, as shown later in Table 5.3, there is not much difference between the association and non-association members regarding social insurance enrolment and health insurance coverage among the sector workers.

Transport workers in both the matatu and motorcycle sub-sectors have strong disruptive power which they exercise collectively as workers at different levels to disrupt the economy and demand for their rights. Though not formally documented, the informal transport sector contributes greatly to Kenya's economy and is a key employer. Matatu and bodaboda operators are many and have strong power which they use from time to time to make demands on issues that affect them. For instance, they use blockages, strikes, threats, and traffic disruptions to demand for action from government and authorities. There have been instances where matatu or bodaboda workers collectively, but separately, use regime-disruptive power to resist government directives, demand for leniency, or push the government to retract decisions that threaten their jobs or sector operations. Recent examples include paralyzing of transport operations in Nairobi due to poor state of roads, introduction of the bus-rapid transit (BRT), intended removal of matatus from the central business district, increased fuel prices, or police harassment and crackdowns. The bodaboda operators also block streets to

Table 5.3 Differences in key workers characteristics by association member status

	Member	*Not Member*	*Difference*	*t-Value*
Social insurance enrolment	0.49	0.44	0.05	0.67
Health insurance coverage	0.52	0.45	0.06	0.79
Nairobi	0.55	0.65	−0.10	-1.31
Kisumu	0.45	0.35	0.10	1.31
Gender (male = 1)	0.96	1.00	−0.04	-1.53
Married	0.88	0.82	0.06	1.19
Local	0.30	0.27	0.02	0.33
Age	36.28	35.85	0.42	0.30
Mean daily earnings (current USD)	10.88	11.16	−0.28	−0.25
Assets (house and/or land)	0.21	0.44	−0.22	-3.21***
Primary incomplete	0.11	0.09	0.02	0.40
Primary complete	0.47	0.49	−0.02	−0.28
Secondary or above	0.42	0.42	0.00	0.03
Training course	0.43	0.49	0.06	−0.80
Training on job	0.27	0.27	0.00	0.05
Self-taught	0.30	0.24	−0.07	−0.94
Wage-worker	0.63	0.75	0.12	1.57
Own-account	0.30	0.24	−0.07	−0.94
Micro-business	0.07	0.02	−0.05	-1.41
Observations	200			

Source: Author's elaboration based on the project survey data.
Notes: <0.01***, <0.05**, <0.1*.

demand for release of arrested peers and to be allowed access to the CBD. Compared to matatu workers, the motorcycle workers have less recognition but have higher disruptive power. They are perceived more negatively by the public compared to the matatu workers. Hence, their disruptive power is targeted not only to the authorities but also to the public, for instance, when accidents involving one of their own occurs. Incidences of burning of vehicles involved in accidents with bodabodas have been common across the country forcing the government to issue stern warning to the operators.

Associational power

The informal sector is known to have strong associational power which has, over the years, been leveraged by various agencies to economically empower and organize workers in the sector. In Kenya, the matatu sector workers belong to transport SACCOs that are regulated by government and to informal groups commonly known as *chamas*. Bodaboda operators also belong to self-regulating associations and informal groups. The SACCOs, associations, and *chamas* provide varying support to advance owners' and workers' interests, and this depends on the reasons for formation. In the case of matatu SACCOs, workers join them to be registered with NTSA and to access jobs. In some cases, the SACCOs, the bodaboda associations, and even some *chamas* facilitate workers' access to saving facilities and access to loans. Some also support workers' welfare needs. All-in-all, the intentions and reasons for joining the associations vary. Joining the SACCOs is mandatory for matatu workers, but this is not a requirement for bodaboda operators. Also, unlike the bodaboda associations that are formed voluntarily by sub-sector operators, matatu SACCOs are formed mainly by vehicle owners, and the workers fit in and are not involved in governance of the SACCOs. Hence matatu workers have little allegiance to their SACCOs and doubt the SACCOs' willingness to protect workers' interests. This is unlike the bodaboda operators who have greater attachment to their associations but, as noted earlier, are shakier due to internal and external interference.

Institutional power

Institutional power has to do with laws, regulations, procedures, and practices that regulate relationships between workers, employers, and authorities. In Kenya, institutional power exercised over the matatu sector is strong and is often used to regulate operators. There are formal institutions that regulate the sector, like the NTSA and the NPS, but these do not necessarily translate into power for the workers. This chapter maps the larger institutional framework and further reflects on the position of the workers within this framework. Results show that, to work in the sector, one must belong to a registered transport SACCO. Therefore, by default, joining of SACCOs is institutionally pre-determined. The NTSA requires SACCOs and employers to hire workers on formal contracts and provide them with social security. In practice, though, this does not happen, and as noted earlier, most matatu workers are hired by vehicles owners and not the SACCOs, mainly on individually negotiated verbal work contracts. A few SACCOs like KMO and Olokise SACCOs in Nairobi and Kihomi SACCO in Kisumu provide work contracts and social security for workers. However, most SACCOs do not engage workers formally partly due to reluctance by vehicle owners to fully release the vehicles to SACCOs to manage them. Thus, whereas there are legal provisions for employment of workers, in practice, these provisions are not enforced. Hence, the institutional power is exercised at the level of control by the government over the sector, and by SACCOs, over the workers while serving the interests of vehicle owners.

The situation is however different for motorcycle operators. First, it is not mandatory for them to belong to SACCOs. The government has also not provided regulations for hiring bodaboda workers, even though a higher number are own-account operators. Nonetheless, most of them belong to work-related associations that are registered under the Associations Act. Once registered, the associations can enrol workers and also decide on the kind of support to provide to them.

The issuance of PSV licenses by NTSA to SACCOs and badges to workers is a form of control and regulation of the sector. Only licensed vehicles and workers can work in the sector, even though there are some workers who find their way into the sector even when not licensed like the reliever drivers, known as squad drivers, who work on mutually agreed terms with the regular drivers towards the end of trips. The licensing requirement nonetheless has protective benefits for matatu workers as only authorized persons can work in the sector. The benefits are even greater for the drivers given that only those having Class A driving license are permitted to drive PSVs. This raises their competitive value (a form of marketplace bargaining power) and makes them more difficult to replace, thus increasing their marketplace bargaining power discussed above. Discussions with the workers revealed that Class A drivers are highly sought after. There is nonetheless a negative effect associated with high staff turn-over due to the common practice of poaching drivers, especially by new vehicle owners. This further weakens the SACCOs ability to contain workers and to negotiate their access to social security. An official of Kihomi SACCO noted that, annually, about 12 out of 52 drivers and 20 out of 30 conductors leave their jobs despite having formal work contracts and regular salary. In general, SACCOs cannot restrict workers' job mobility unless they have unpaid SACCO loans or pending criminal cases that hold SACCOs accountable.

Figure 5.1 summarizes the authors' interpretation of the relationships between the government, SACCOs, vehicle owners, and workers in the matatu sub-sector. The relationships, though co-dependent especially between SACCOs, owners, and workers, are also based on a power hierarchy where the government wields overall power, whereas the SACCOs and owners have power over workers. However, even in this arrangement, the workers hold some financial control over the owners since they collect fares and remit daily to the owners as per agreed target.

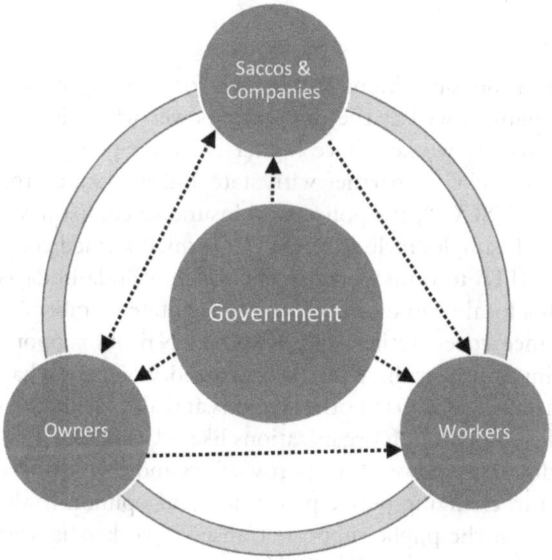

Figure 5.1 Power relations in the matatu sub-sector.

The scenario within the matatu sector is one of mutual dependency. SACCOs depend on owners to give them vehicles, so as to be registered with NTSA. The SACCOs support the owners in registration and licensing of vehicles and the workers by NTSA. Vehicle owners are the employers, even though, ideally, it is the SACCOs that should employ workers. Few vehicle owners agree to franchise their vehicles to SACCOs to manage them, and this reduces the SACCOs authority over workers. In practice, the SACCOs largely favour the interest of vehicle owners and pay little attention to workers' issues. Hence, there are tensions between SACCOs and workers. Associations like MOA and KMO mainly protect vehicle owners' interests. Others like MWA, MaCREW, and Kisumu Boda-boda Association engage with the government and authorities to pursue the SACCOs, associations, and workers' interests and long-term policy changes in the sector. For example, MWA successfully lobbied for reduction of insurance premiums for 14-seater matatus, and this benefited the entire sector. Also, in 2018, 31 SACCOs petitioned the government to suspend the removal of 14-seater PSVs from operating in Kenya (ROK, 2018). There are also transport unions like TAWU, MWU, and PUTON[3] that cater for workers' interests. However, none of the workers in the survey belonged to a union, but other studies in Kenya have covered them (Wright, 2018; Spooner & Manga, 2019). The low participation in union activities is linked to limited time among workers to participate in union activities due to nature of their work, or their SACCOs and employers dissapproval of their joining unions. Some of the unions like TAWU were established to cater for formally employed workers, and it has only very recently begun to recruit informal workers, whereas the independent union MWU was established specifically to cater for informal workers.

Societal power

Societal power includes coalitions that involve forming alliances with actors and other social groups as well as the discursive power which is about influencing the public. Transport workers have higher coalition power through their associations. Some SACCOs partner with state and non-state organizations like NTSA, NITA, KUSCCO, the police, and insurance companies to support and benefit workers. Examples include KAMTCO which formed coalitions with the BRITAM and NITA to train workers. The Kisumu Boda-boda association also partnered with a local insurance company to facilitate members' access to training and insurance cover. Others like KMO in Nairobi supported workers to access health insurance, while Kihomi organized workers exchange visits with successful transport SACCOs in other regions and also facilitated workers access to savings and credit through organizations like KUSCCO.

Regarding discursive power, transport workers and their associations have not achieved much in changing public perception and opinion towards the sector. The only time that the public supports transport workers is when they are directly affected, for instance, when PSVs are barred from accessing the CBD. Often, the public is on the receiving end when transport workers conflict with authorities, for instance, when fuel costs are increased leading to increase in fares. The perception is even worse with regard to motorcycle operators due to poor accident ratings, and their tendency to mob and harass motorists involved in accidents with one of their own. Thus, even when workers are disadvantaged or denied their right to social protection, public opinion is unlikely to tilt in favour of workers.

Discussion

Comparison between members and non-members

Table 5.3 shows that there are no significant differences along a variety of dimensions between association members and non-members, except for asset ownership which is higher among the non-members. Non-members also have higher earnings and are more likely to be in Nairobi (although the differences are not statistically significant). Members consist of a relatively higher proportion of own-account operators and a lower proportion of wage-workers compared to non-members. This is likely due to the more insecure nature of the former work type, in turn, increasing the need for associational belonging.

Access to formal social insurance

Despite the government requirement that transport workers should have social insurance, around 50 percent of workers remain uncovered as seen in Tables 5.1 and 5.2. Tables 5.3 and 5.4 reveal that, although association members in general are more likely to have coverage, the difference is not statistically significant, with the exception of SACCO members, as discussed further on. This could be due to the fact that, on the whole, the vehicle owners expect the workers to pay for their own coverage due to the notion that they control the daily incomes. In the matatu sector, fare payment is largely cash and is often handled by workers. Alternative cashless payment options are not common, and this makes the owners suspicious that the workers do not remit the entire daily earnings. To deal with this fear, workers are put on a set daily target (Spooner & Manga, 2019) and are left to cater for their social security needs. Unlike the formal sector, where part of the contributions, for instance, to pension coverage, is paid by the employer, this is not the case in the informal sector – including even those who have formal work contracts from SACCOs. As noted earlier, workers who are enrolled in social security programmes through the SACCOs still pay for their own coverage. Table 5.4 also shows that male and self-taught workers are less likely to have formal social insurance; yet, the sector is male-dominated.

Factors that contribute to workers' low social insurance include employers' refusal to pay for workers' coverage, high worker turn-over, workers' unwillingness to enrol in employer-linked schemes, low awareness, and high cost of premiums. A few SACCOs comply with NTSA requirements to enrol workers

Table 5.4 Differences in key workers characteristics by social insurance enrolment

	Social Insurance	*No Social Insurance*	*Difference*	*t-Value*
Association member	0.75	0.70	0.04	0.67
Sacco/vicoba/chama	0.65	0.51	0.14	1.99**
Work-related association	0.08	0.15	-0.07	-1.48
Women/youth/religious	0.01	0.04	-0.03	-1.25
Work-related	0.29	0.30	0.00	0.01
Loans	0.36	0.31	0.04	0.65
Voice and representation	0.05	0.08	-0.02	-0.67
Social cushioning	0.04	0.02	0.02	0.95
Kisumu	0.63	0.53	0.10	1.41
Nairobi	0.37	0.47	-0.10	-1.41
Gender (male = 1)	0.95	0.99	-0.04	-1.79*
Married	0.89	0.84	0.06	1.17
Local	0.27	0.30	-0.03	-0.48
Age	36.99	35.41	1.58	1.26
Mean daily earnings (current USD)	11.02	10.90	0.12	0.12
Assets (house and/or land)	0.22	0.32	-0.10	-1.63
Primary incomplete	0.08	0.12	-0.04	-0.91
Primary complete	0.46	0.49	-0.02	-0.32
Secondary or above	0.45	0.39	0.06	0.89
Training course	0.47	0.42	0.05	0.77
Training on job	0.31	0.24	0.07	1.07
Self-taught	0.22	0.34	-0.12	-1.91*
Wage-worker	0.64	0.68	-0.03	-0.51
Own-account	0.29	0.28	0.02	0.29
Micro-business	0.06	0.05	0.02	0.48
Observations	200			

Source: Author's elaboration based on the project survey data.
Notes: <0.01***, <0.05**, <0.1*.

in social insurance while others encourage the workers to enrol voluntarily. This could explain the finding that SACCO members are more likely to be enrolled in social insurance (Table 5.4). However, the reluctance among vehicle owners to allow SACCOs to fully manage their vehicles including employing workers weakens the SACCOs ability to ensure workers' provision of social insurance. Hence, the workers' access to social insurance is not dependent on association membership but on whether the SACCOs they belong to have full mandate to manage worker issues.

The co-dependence between SACCOs and vehicle owners and lack of structures to ensure compliance with government directives compromises workers' access to social protection. The lack of full control of incomes from the vehicles by the owners makes them reluctant to use their earnings, which they consider to be low, to provide for workers' social insurance. The owners' perspective is that workers control the earnings from the vehicles and that they can afford to pay for their own social security coverage. Hence, many employers do not consider provision of workers' social protection as their responsibility. The workers on their part argue that their net incomes are low and that they have competing personal and family demands and therefore not in a position to pay for social security. In Kisumu, the vehicle owners observed that:

> The workers are our bosses; they employ us because they control the amount that the vehicles earn. They determine how much money the vehicle owners get at the end of the day, after they have paid themselves. They have a lot of money and they can afford to pay for their social protection coverage. The vehicle owners have little control of the vehicle operations and there are no systems to protect us. The money that we get is mainly used to repay the vehicle loans.
>
> (FGD, Kisumu)

Conclusion

The informal public transport sector is one of the largest in Kenya. It includes matatu and motorcycle sub-sectors. The sector is highly regulated by government. For instance, it is mandatory for matatu workers to belong to SACCOs. This is, however, not mandatory for motorcycle operators, even though most of them belong to sector-based associations. Transport SACCOs and associations benefit workers by engaging with and lobbying authorities on behalf of workers on operations-related issues. However, they hardly address workers' social protection issues like social insurance provision, even when this is required by NTSA in the case of matatus. The sector is beset with many challenges that affect workers' social protection provision, like lack of formal work contracts. Consequently, most workers depend on informal systems and networks when faced with shocks. The reluctance by vehicle owners to franchise vehicles to SACCOs for management, lack of enforcement to compel employers to provide workers with formal contracts and social security, and workers' fear of joining labour unions contribute to the low social security coverage among workers. This also denies workers their right to work in favourable and decent work environments.

Whereas transport SACCOs and associations exist, they mainly protect the interests of vehicle owners and not of workers. Hence, the workers fear to voice their concerns or to approach their associations whenever they have work-related challenges, except when they require the SACCOs and associations to engage with authorities on their behalf. Some SACCOs support workers by providing training, access to credit, or catering for welfare needs. There is, however, little variation in social protection coverage between workers who belong to associations and those who do not. This chapter thus concludes that having government social protection policies and regulations in place is important. However, this does not guarantee workers access to social protection. The role of transport SACCOs and associations needs to be re-examined with a view to compelling them to focus on workers' issues including their right to social insurance.

Notes

1 Workers in Kisumu indicated that MOA was inactive there and that not all owners are members.
2 https://www.2nksacco.co.ke/.
3 PUTON is an independent democratic registered trade union in Kenya which is affiliated to the International Transport Workers Federation. It draws membership and represents workers in transport-oriented occupations.

References

Behrens, R., McCormick, D., Orero, R. & Ommeh, M. (2017) Improving paratransit service: Lessons from inter-city matatu cooperatives in Kenya. *Transport Policy.* 53, 79–88. Available from: doi:10.1016/j.tranpol.2016.09.003.

Devereux, S. & Sabates-Wheeler, R. (2004) *Transformative social protection.* IDS Working Paper 232. Brighton, Institute of Development Studies.

ILO (2020) *Case studies: Developing a business case for labour law compliance.* International Labour Organization. Geneva, International Labour Office. Available from: https://www.ilo.org/wcmsp5/groups/public/---africa/---ro-abidjan/---ilo-dar_es_salaam/documents/genericdocument/wcms_500927.pdf.

Kamau, A. & Mitullah, W.V. (forthcoming) Women's value creation in informal public transport enterprises in Kenya. In: Yousafzai, S., Henry, C., Boddington, M., Sheikh, S. & Fayolle, A. (eds.) *Research handbook of women's entrepreneurship and value creation.* Cheltenham, UK, Edward Elgar Publishing Limited.

Kamau, A., Kamau, P., Muia, D., Baiya, H. & and Ndung'u, J. (2018) Bridging entrepreneurial gender gap through social protection among women small scale traders in Kenya. In: Yousafzai, S., Fayolle, A., Lindgreen, A., Henry, C., Saeed, S. & Sheikh, S. (eds.) *Women's entrepreneurship and the myth of 'underperformance'.* Cheltenham, UK: Edward Elgar Publishing Limited.

Kamau, A., Michuki, G., Kamau, P. & Mwangi, S. (2019) *Overcoming challenges in extending pension coverage to informal sector workers in Kenya.* Research Paper, Retirement Benefits Authority (RBA).

Lund, F. (2009) Social protection and the informal economy: Linkages and good practices for poverty reduction and empowerment. In: *Promoting pro-poor growth: Social protection.* Geneva, Organisation for Economic Cooperation and Development (OECD), pp. 69–88.

McClean, E.J., Burris, E.R. & Detert, J.R. (2013) When does voice lead to exit? It depends on leadership. *The Academy of Management Journal.* 56 (2), 525–548. Available from: http://www.jstor.com/stable/23412601.

Mutongi, K. (2017) *Matatu: A history of popular transportation in Nairobi.* Chicago, IL, University of Chicago Press.

Mwanza, E. (10 February 2021) Nairobi residents rank best matatu Saccos. *Kenyans.co.ke.* Available from: https://www.kenyans.co.ke/news/62090-nairobi-residents-rank-best-matatu-saccos.

Mwobobia, B. (2011) *Critical success factors in the motorcycle boda-boda business in Nairobi, Kenya.* MA thesis. University of Nairobi.

NTSA (2014) *National transport and safety authority (operation of public service vehicles) regulations, 2014: Subsidiary legislation.* National Transport and Safety Authority. Nairobi, Government Printer.

Omulo, C. (12 February 2020) Revealed: 420 matatu saccos operating in Nairobi illegally. *Daily Nation*. Available from: https://www.nation.co.ke/kenya/counties/nairobi/revealed-420-matatu-saccos-operating-in-nairobi-illegally--249852.

Opondo, V. & Kiprop, G. (2018) *Boda motorcycle transport and security challenges in Kenya*. NCRC Research Report No. 14. National Crime Research Centre. Nairobi, The Jomo Kenyatta Foundation.

ROK (2012a) *Laws of Kenya: Transport licensing act chapter 404*. Revised Edition 2012 [1979]. Republic of Kenya, National Council for Law Reporting with the Authority of the Attorney-General.

ROK (2012b) *Laws of Kenya: Co-operative societies act chapter 490*. Revised Edition 2012 [2005]. Republic of Kenya, National Council for Law Reporting with the Authority of the Attorney-General.

ROK (2018) *In the high court of Kenya at Nairobi constitutional & human rights division petition no. 440 of 2018*. Republic of Kenya. Available from: http://kenyalaw.org/caselaw/cases/view/198871/.

ROK (2020) *The traffic (driving schools, driving instructors and driving licences) rules, 2020*. Legal Notice No. 28 The Traffic Act (Cap. 403). Special Issue Kenya Gazette Supplement No. 19 (Legislative Supplement No. 12), 10 March 2020. Available from: http://kenyalaw.org/kl/fileadmin/pdfdownloads/LegalNotices/2020/LN28_2020.pdf.

Spooner, D. & Manga, E. (2019) *Nairobi bus rapid transit: Labour impact assessment research report*. Manchester, Global Labour Institute. Available from: https://www.researchgate.net/publication/330398422_Nairobi_Bus_Rapid_Transit_Labour_Impact_Assessment_Research_Report.

Spooner, D. & Whelligan, J. (2017) *The Power of Informal Transport Workers: An ITF Education Booklet*. Manchester, Global Labour Institute. Available from: https://www.itfglobal.org/media/1691170/informal-transport-workers.pdf.

World Bank (2019) *Implementation completion and results report: Report no: ICR00004798*. Available from: http://documents1.worldbank.org/curated/en/702181563299068935/pdf/Kenya-National-Urban-Transport-Improvement-Project.pdf.

Wright, T. (2018) *The impact of the future of work for women in public transport*. International Transport Workers' Federation (ITF), Friedrich Ebert Stiftung & Women Transporting the world.

6 Informal trader associations in Tanzania – providing limited but much-needed informal social protection

Lone Riisgaard

Introduction

Micro-trade is a dominant activity in the informal economy in most urban areas of the global south. In Tanzania, it provides a very relevant context for studying the access that people in the informal economy have to both formal and informal social protection mechanisms – insights that are essential in light of the ongoing expansion of formal social protection in Tanzania and Sub-Saharan Africa more broadly. As described in more detail in the introduction to this edited volume, social protection policies and instruments have gained increasing attention among many countries, including Tanzania. Many partners, including donors, have since the late 1980 viewed social protection as the preferred solution to issues of poverty in the global South (Deacon, 2013; Hickey & Seekings, 2017; Hickey et al., 2019).

In Tanzania, recent years have seen attempts at expanding informal workers' access to social protection such as pension and insurance schemes. Nonetheless, as illustrated in this chapter, formal social protection is still largely beyond the reach of informal traders, a gap which informal traders' associations seek to fill with a variety of different services. They, for example, cushion members in cases of death or medical problems in the family and represent members' interests towards authorities.

In this chapter, we investigate the access that informal micro-traders in Tanzania have to formal and informal social protection measures. The key challenges they face and how (and if) these challenges are addressed by formal social protection measures and by the informal social protection measures employed by their own collective associations are discussed. Consequently, the chapter provides insights into how micro-traders – who work in one of the most common and vulnerable informal occupations – organize and how (if at all) they access social protection. By comparing Dodoma's setting with Dar es Salaam, which is predominantly urban, we also seek to address differences (if any exist) which might be related to the degree of urbanization.

A broad understanding of social protection is adopted, encompassing formal social insurance programmes such as health insurance or maternity leave as well as informal cushioning mechanisms, access to micro-finance and training, and representation and voice. However, narrowly targeted safety net measures

DOI: 10.4324/9781003173694-6

(commonly known as social assistance) are not covered since they are mostly not offered by informal associations.

The chapter employs a broad understanding of the informal economy, including self-employment in informal enterprises (i.e. unregistered or unlicenced business), as well as wage employment in informal jobs without a written contract. As discussed in more detail in Chapter 1 of this edited volume, informality is a highly politicized concept which often appears in degrees while intersecting with formality in various ways. Informal traders might, for example, sell goods on commission for formally registered retailers while working either in formally designated vending areas or in informal ones. Nonetheless, informality remains an essential concept in terms of delimitating social protection rights for workers.

The chapter draws on Nvivo-processed data from focused group discussions (FDGs) and key informant interviews (KIIs) with micro-traders and their associations carried out between April 2018 and January 2019.[1] Four FGDs were conducted, representing a total of 13 different associations in Dodoma and 16 in Dar es Salaam. Furthermore, nine KIIs were conducted with association leaders in Dodoma and 12 in Dar es Salaam. In total, more than 40 different associations were represented. The paper also draws on survey data on micro-traders in Dodoma and Dar es Salaam. The survey was carried out in 2018 and used a combination of purpose-based sampling and random selection. Of the 286 survey interviews conducted with micro-traders, 62 percent were sampled geographically while 38 percent were sampled via a variety of informal worker associations (where the target group were ordinary members) to ensure a broad coverage of different types of associations.[2]

The findings are indicative of the collective organization and access to social protection amongst micro-traders in urban areas and, in particular, the more vulnerable segment of micro-traders. We focused on traders who were mobile and/or working in a space without a fixed structure even though a few KII and FGD participants did come from established markets. Most often, their work is characterized by low entry barriers, high competition, and low and irregular incomes. They generally have no officially recognized space of work and face frequent harassment in environments which lack essential facilities such as clean water, waste disposal, and storage facilities.

In the remainder of this chapter, we start by examining how informal trade is governed in Tanzania, and relatedly, how informal traders are organizing collectively; we then explore the challenges faced by micro-traders and their access to formal social protection mechanisms; followed by discussion of how micro-traders are organized and the services associations offer to their members. Using the Power Resource Approach (PRA; Schmalz, Ludwig & Webster, 2018), the power resources available to associations are analyzed along with their access barriers. We then compare traders who enrol in informal associations to those who do not belong to any associations. The final section concludes on the findings and the overall comparison between informal social protection mechanisms and official ones.

Collective organization and the governance of micro-traders

Attitudes and policies towards informal micro-traders have varied over time. In Dar es Salaam, local by-laws from the 1960s and 1970s made micro-trading illegal, but the 1980s economic crisis resulted in some accommodation such as the establishment of municipal licences for mobile traders in 1983 (Brown, Lyons & Dankoco, 2009; Brown et al., 2015). In general, a more participatory approach to city planning was initiated where the local government in Dar es Salaam encouraged representation of informal workers' associations at the ward level, and by 1997, some 240 self-help groups had been formed (Brown, Lyons & Dankoco, 2009).

Since 2003, however, the participatory line has been reversed. In 2003, the peddlers' licence was cancelled, rendering street traders illegal (Lyons & Msoka, 2010), and in 2006, the Prime Minister instructed local authorities to move street traders from busy areas, leading to widespread clearances (Lyons & Msoka, 2010). Evictions have been and still are common, although often relocation has been unsuccessful as the new localities have failed to attract customers.

Tanzania has a long tradition of traders' associations and other informal networks which was intensified by the post-independence norms of socialism and self-reliance (Brown, Lyons & Dankoco, 2009). Associations are commonly known by the Swahili word kikundi (plural: vikundi) and are often welfare-oriented, multi-ethnic, run by contributions from members and have elected officers (Tsuruta, 2006).

In addition to trade associations are the trade unions, which historically have been relatively weak in Tanzania, as the ruling party has sought to control independent sources of power (Tripp, 2000). The autonomy of the trade unions increased with the reintroduction of a multi-party system in the early 1990s; nonetheless, trade unions in Tanzania are still swaying between continued control by the ruling party and efforts to establish themselves as autonomous organizations (Fisher, 2011).

The trade union movement in Tanzania has only within the last decade or so begun to show an interest in organizing informal workers (Fisher, 2011). In a context of declining trade union membership, trade unions in Tanzania are taking new measures to include people in the informal economy as part of the trade union constituency. Lacking any strategy on the matter from the federation level, very few unions are yet to succeed in recruiting substantial numbers of informal members. Only the unions representing transport workers, domestic workers, and commercial workers are actively pursuing membership of informal workers. This is often in cooperation with labour-oriented NGOs and international or developed country unions which now have inclusion of the informal economy as a top priority.

Since the late 1990s, governments have tried to formalize informal activities with the aim of enabling small informal enterprises to participate in and benefit from the market economy (Brown, Lyons & Dankoco, 2009). An example is the 2006 World Bank-supported 'Doing Business' programme. It assumed that micro-entrepreneurs such as informal traders would benefit from reduced bureaucracy and in particular to property and business rights and increased access to credit (Lyons, Brown & Msoka, 2014). However, as shown by Lyons, Brown, and Msoka (2014), very few licence categories have been available for micro-traders and to obtain a business licence, one needs to have a fixed and legal address. Hence, the requirements demanded to formalize cannot be met by most micro-traders.

The strong focus on formalization gives the impression that informal activities are undesired and something to be overcome in order to achieve a modern market-driven economy. Discourses, however, change, and the late president Magufuli advocated a stop to harassment of informal traders. Hence, many small traders were able to paraphrase the president as illustrated by one street vendor: "The president said no eviction of these people. Leave them to work. You should make arrangement of how they will work in the streets" (FGD, Dar es Salaam). The more positive attitude expressed by the president has gone together with a massive effort of registration of individual traders, along with what resembles a new form of taxation.

Since December 2017, a total of 670,000 IDs (popularly called Magufuli cards) have been distributed to micro-traders with business capital not exceeding four million TZS (1,769 USD) per annum (Tanzania Daily News 02.01.2018, The East African 15.12.2018). The total value of the IDs (at 20,000 TZS each/9 USD) is 13.4 billion TZS (6.03 million USD).[3] All revenue from the new ID cards goes to the central government, and hence they have been criticized for diverting revenue away from the local governments who are still responsible for servicing the areas where the micro-traders work – but are no longer supposed to collect trading fees.[4] The new ID so far seems to have been generally well received by micro-traders, although it is somewhat unclear what they get from having this card.

In sum, authorities have had, and still have, an ambivalent relation with informal traders. Approaches alternate between tolerance and evictions, often following election cycles, as politicians ask local authorities to go easy on informal traders until the election is over (see e.g. Babere, 2013). Despite the new IDs, micro-traders have very few rights in public space, and if found guilty of trading in unsanctioned areas, they face a fine of TZS 50,000 (USD 21.90).[5] Meanwhile, efforts at formalizing the informal economy – at the moment predominantly via registration and relocation – continue.

Characteristics of micro-traders

The micro-traders in Dar es Salaam and Dodoma form a heterogeneous group. They all provide cheap goods and service but occupy very different positions in the economy depending on what they sell and whether they work in designated or undesignated areas. Of the 286 micro-traders participating in the survey, the average age was 35 years, and 56 percent were female. The educational level varies quite a lot but, perhaps tellingly, less than a quarter had above primary level education (Table 6.1). These figures are in line with those reported for the informal economy in the 2014 Labour Force Survey (ILFS, 2015).

The heterogeneity of informal traders is illustrated in the assets they own. As seen in Table 6.1, 42 percent of the survey respondents (or their spouse) own a plot of land or property while 15 percent own other assets.[6] An equal amount (43 percent), however, do not possess any assets. The median daily wage listed in Table 6.1 also hides large differences. As seen in Table 6.2 below, there is a large difference between the 10th and the 90th percentiles (1 USD versus 17.5 USD daily); however, 75 percent earn 7.66 USD or less per day.

Table 6.1 Key worker characteristics

	Tanzania N = 286			*Dar es Salaam N = 134*		*Dodoma N = 152*	
	Sum	*Mean*	*Std. Deviation*	*Mean*	*Std. Deviation*	*Mean*	*Std. Deviation*
Association member*		0.34	0.47	0.33	0.42	0.34	0.48
Formal insurance enrolment (health/ pension)	57	0.20	0.40	0.09	0.29	0.30	0.46
Health insurance coverage	64	0.22	0.42	0.13	0.33	0.31	0.46
Gender (male = 1)	127	0.44	0.50	0.53	0.50	0.37	0.48
Age		34.85	11.19	36.04	11.82	33.8	10.53
Married	162	0.57	0.50	0.55	0.50	0.58	0.50
Local born	134	0.47	0.50	0.25	0.43	0.66	0.48
Mean daily earnings (current USD)**		7.87	13.83	8.46	12.15	7.36	15.06
Assets (house and/or land)	119	0.42	0.50	0.34	0.48	0.48	0.50
Primary incomplete	38	0.13	0.34	0.1	0.30	0.16	0.37
Primary complete	183	0.64	0.48	0.69	0.47	0.6	0.49
Secondary or above	65	0.23	0.42	0.22	0.41	0.24	0.43
Wage-worker	16	0.06	0.23	0.05	0.22	0.06	0.24
Own-account	223	0.78	0.42	0.82	0.39	0.74	0.44
Micro-business	46	0.16	0.37	0.13	0.33	0.19	0.39

Source: Author's elaboration based on project survey data.

Notes: Summary statistics based on the full sample. * Association member figures are from the random sample where N = 179, N = 89, and N = 90 respectively. ** Due to the large standard deviation and the large difference between the mean and median values, the median value 3.72 USD is thought to be a more accurate indication of the 'general' earning level. The median value for Dar es Salaam is USD 5.38 and for Dodoma it is USD 3.28.

Table 6.2 Percentiles for daily earnings

10%	*25%*	*50%*	*75%*	*90%*
USD 0.99 TZS 2,262	USD 1.97 TZS 4,500	USD 3.72 TZS 8,500	USD 7.66 TZS 17,500	USD 17.51 TZS 40,000

Source: Author's elaboration based on project survey data.

Additionally, large differences between "highest" and "lowest" take-home earnings indicate highly fluctuating incomes.[7]

Micro-trade was the main source of income for the vast majority (94 percent), although 29 percent had a secondary occupation. The largest group (40 percent) were street vendors selling different items such as magazines or sweets, 35 percent were selling cooked foods or snacks, and 15 percent different food items such as fruits or grains. The remaining 10 percent were involved in different service activities such as car washing or shoe shining.

Employment-related benefits like paid maternity leave or work-injury compensation were practically non-existent. This is not surprising, as most micro-traders have no direct employers – only 6 percent consider themselves wage-workers. The majority are own-account workers (78 percent) or micro-enterprise owners with one or two helpers (16 percent). Even though some traders work on commission for larger retailers, they are not in any standard employment relationship. In Tanzanian labour legislation, employment-related social protection is linked to a standard employment relationship and hence does not cover micro-traders. This means that micro-traders have to sign up voluntarily and make the full contribution out of their own pockets. Relatedly, as seen from Table 6.1, social insurance enrolment and coverage in the areas of health and pension only cover 20 percent and 22 percent, respectively.

For most of the key characteristics illustrated in Table 6.1, differences between Dar es Salaam and Dodoma were minor, indicating consistency across locations for issues such as educational level or age. The gender profile, however, differed slightly as the sample consisted of 47 percent females in Dar es Salaam but 69 percent in Dodoma. Median wages were lower in Dodoma (3.28 USD) compared to Dar es Salaam (5.38 USD), and similarly, a larger proportion of traders in Dar es Salaam were immigrants to the city (75 percent), whereas, in Dodoma, 66 percent were locally born. This might explain why more traders in Dodoma held key assets (land or property) in their family than in Dar es Salaam (48 percent versus 34 percent). Finally, social insurance coverage was markedly higher in Dodoma (31 percent) compared to Dar es Salaam (13 percent) which can be explained by the prevalence of the Community Health Fund (CHF) outside the capital as discussed below.

Access to formal insurance schemes

Tanzania has three public insurance schemes potentially encompassing informal operators. That is the National Social Security Fund (NSSF), the National

Health Insurance Fund (NHIF), and the CHF which has recently been re-launched as the improved Community health fund (iCHF) under the NHIF (and co-financed by the government) which targets rural populations and people without formal employment.

The survey shows that enrolment in pension schemes is very rare amongst micro-traders as only 1 percent reported participation. This was substantiated by interviews where respondents considered pensions to be out of their reach. The criteria that one has to contribute for 15 years to be eligible for pension under the current NSSF scheme makes it fit poorly with most informal traders who tend to focus more on short term needs.

A total of 19 percent of the micro-traders contributed to official health insurance divided between CHF (12 percent), NHIF (5 percent), NSSF (1 percent), and other insurance (1 percent). Actual coverage (including cases where a spouse or relative is the primary contributor) shows 22 percent coverage with a similar division between the different schemes. This corresponds to the national coverage of 22 percent estimated in Jacob and Pedersen (2018).

Thirteen percent of the respondents who were not currently contributing to an NHIF scheme had previously done so. Common reasons given for exiting were related to leaving a formal job, the quality of the service or the cost. Amongst the 229 persons from the survey who did not contribute, the most common answer was "it is too expensive", followed by "no knowledge about insurance" and "procedures are too complicated". Hence, problems related to costs, procedures, information flows, and retainment inhibit coverage of micro-traders in the existing schemes.

What characterizes micro-traders enrolled in formal insurance schemes compared to non-enrolled traders?

In order to examine the relations between formal social insurance enrolment and key worker characteristics, we elaborate a test of means (see Table 6.3). Table 6.3 indicates that enrolled traders are more likely than non-enrolled to be members of an association, to live in Dodoma, be female, married, own key assets, have higher education, and to be either own-account or micro-businesses. Nonetheless, when testing these relations using a standard probit model, only the variables assets, secondary education, and location remain significant (see Riisgaard, 2020).

This leads to the conclusion that traders living in Dodoma are more likely than traders living in Dar es Salaam to enrol in formal insurance schemes and that, on average, traders with a house or a piece of land in the family are also more likely to enrol compared to ones without these assets. Finally, traders with secondary education or above are much more likely to enrol compared to traders who have not completed primary education. These findings seem to indicate that official health insurance mainly caters for the segment of informal traders who are already more resilient (in that they are more likely to have assets in the family and have higher education) but also that access to formal health insurance

Table 6.3 Differences in key workers characteristics by social insurance enrolment

	Enrolled Mean	*Non-enrolled Mean*	*Mean Difference*	*T*
Association member	0.68	0.55	0.13	1.78*
Sacco/vicoba/kikundi	0.53	0.33	0.20	2.82***
Work-related association	0.11	0.19	-0.09	-1.55
Women/youth/religious	0.05	0.03	0.012	0.62
Work-related benefit	0.19	0.17	0.023	0.48
Social cushioning	0.14	0.08	0.06	1.45
Voice and representation	0.02	0.04	-0.02	-0.80
Loans	0.33	0.27	0.06	0.94
Dar es Salaam	0.21	0.53	-0.32	-4.50***
Dodoma	0.79	0.47	0.32	4.50***
Gender (male = 1)	0.26	0.49	-0.22	-3.11***
Married	0.68	0.54	0.15	2.01**
Local born	0.56	0.45	0.12	1.57
Age	36.98	34.32	2.66	1.61
Mean daily earnings (current USD)	9.90	7.37	2.53	1.24
Assets (house and/or land)	0.61	0.37	0.245	3.45***
Primary incomplete	0.07	0.15	-0.08	-1.56
Primary complete	0.53	0.67	-0.14	-2.00**
Secondary or above	0.4	0.18	0.22	3.62***
Wage-worker	0.04	0.06	-0.03	-0.76
Own-account	0.67	0.81	-0.14	-2.32**
Micro-business	0.3	0.13	0.17	3.20***

Source: Author's elaboration based on the project survey data.
Notes: N for enrolled = 57, N for non-enrolled = 229. *$p < 0.10$, **$p < 0.05$, ***$p < 0.01$.

is more widespread in Dodoma which is most likely related to the cheaper CHF option available there.

During discussions, little interest was expressed in the CHF scheme or its urban counterpart TIKA as the quality of the service offered was considered poor. Of much more interest was the newly introduced KIKOA scheme. In 2015/2016, the government diversified NHIF to engage the informal economy through the MUTUAL (commonly known as KIKOA) scheme. Under this scheme, members of registered associations in the informal economy with a minimum of 20 members can join (if a minimum of 10 members sign up) for 76,800 TZS per year per person. Dependents can be signed up at the cost of an additional 76,800 TZS per person (NHIF, 2018). Several associations had provided access to this scheme for their members, as discussed later in in the section on association services.

Perception of key challenges

When asked during interviews what was perceived to be key challenges, problems with authorities which include evictions, confiscation of goods, fines, harassment, and insecurity of trading spaces were the most mentioned. Lack of access

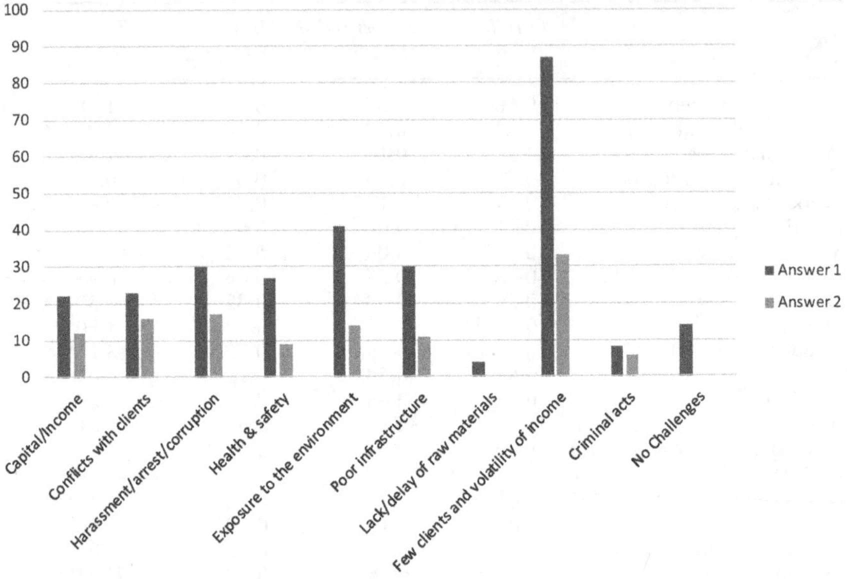

Figure 6.1 Key work-related challenges.
Source: Author's elaboration based on project survey data.

to health insurance, lack of capital, access to credit, and lack of training on business skills were also mentioned as problems.

In the survey, respondents were asked to "Describe the two key challenges/threats/problems you face in your work". As seen in Figure 6.1, a variety of challenges emerged. Issues related to problems with authorities again come out as important although fluctuations in income was the challenge mentioned most often followed by exposure to the environment.

Survey responses and the qualitative interviews indicate a range of work-related challenges affecting informal traders, with the most pertinent challenge varying between individual traders. Of particular importance seems to be limited or fluctuating income (which is also evident in the wide differences between low and high daily income estimates reported), exposure to physical conditions, and problems with authorities as well as lack of access to health insurances. As problems with authorities were perceived as an important challenge in both the qualitative and quantitative data, and since this is also related to many of the other challenges the traders face, below, we elaborate on this aspect based on the qualitative interviews.

Relation to the authorities

Although harassment from authorities was commonly mentioned as a major challenge, there had been some improvements under president Magufuli.

That the president had publicly made statements in favour of street traders, was noted mainly as a welcome sign of appreciation and a shift from being regarded as a nuisance more than it seemed to have actually changed their daily interactions with local authorities. In addition, trust in the longevity of the current, more positive, climate is limited as noted below:

> Although we are out of that mess now since our president Magufuli gave the street vendors the right to perform their activities in Kariakoo market, some of us still do not think that it will be maintained for long so we think maybe someday the system will change and we will go back to the old ways when street vendors were being mistreated and harassed at their business premises.
>
> (KII, Dar es Salaam)

Government actions like evictions can have detrimental effects not just for the individual traders (Skinner, 2008) but also for their associations, as exemplified in the following quote:

> Ok as I said before that when we started our group we used to have a lot of good planning but as I said we didn't survive long since the government come to remove us from there so that action destroy a lot of our planning.
>
> (FGD, Dar es Salaam)

Some associations would seek to assist members in case of problems with authorities though most saw themselves as powerless and incapable of helping their members with what they considered harassment by authorities. As noted by Babere (2013), representation towards authorities is furthermore challenged by the fact that numerous different entities are in charge of governing different aspects of informal trade. In the following, we explore in more detail how associations assist their members.

Micro-traders' associations

Most associations were registered with a constitution guiding their operations. Registration appears to relay a degree of legitimacy in relation to the authorities and in organizing collective support for members such as loans or training. Government promises of loans was frequently mentioned, but rarely materialized. Below, we briefly look at three associations to illustrate some of the variety in terms of organizational set-up and purpose.

VIBINDO society[8]

Adversarial relations between micro-traders and authorities spurred the establishment of VIBINDO in 1995 when 19 associations formed an umbrella organization to represent their interests and negotiate with authorities in attempts to secure trading spaces. VIBINDO has since then offered financial services and training for members, established their own private micro-health insurance scheme, and later negotiated the KIKOA scheme with NHIF.

VIBINDO is organized at the national regional and district level and reports more than 60,000 members organized via 630 groups of small businesses. Their mission is "to assist its members in transforming, formalize and grow their micro-businesses and industries in order to enhance opportunities for access to markets, social security, conducive business premises, financial services ..." (Workshop, Dar es Salaam, May 2018).

The membership is predominantly male (80 percent), and while not limited to micro-traders, this group forms the vast majority. Of the micro-traders, the largest part is own-account, although a few have employees. Hence, VIBINDO has more of a private-sector perspective as compared to, for example, the trade unions that focus on the relationships of employment. They are, for instance, members of Tanzania private-sector foundation (TPSF), a private organization lobbying to promote business interest in national policy platforms such as the Tanzania National Business Council (a public private partnership). In addition, VIBINDO focuses on small enterprises rather than wage-workers.

VIBINDO is a well-known name amongst authorities as well as donors and has had several projects with the ILO. In addition to externally funded projects, their source of funding is membership contributions, in addition to offering commercial services like training and assistance in registration of associations.

TUICO[9]

TUICO has around 80,000 members and is considered one of the better organized and resourced trade unions. In their membership are around 4,000 informal micro-traders and TUICO is an affiliate member of StreetNet International. Their engagement in the informal economy started around 2010 and has been supported by various NGOs.

The branches of informal micro-traders are based in designated markets, as the union has found these informal workers easier to approach. Hence, for practical reasons, access to a somewhat secure vending space is a key consideration for the union when they reach out to informal workers. The implication of this is, however, that they reach the comparatively more secure traders.

A newer initiative from TUICO, originating in an externally funded programme they had with StreetNet International, has been to establish a bargaining committee with two members from each of the market-based branches they have in Dar es Salaam. As noted by a KII in Dar es Salaam, with this initiative, TUICO is seeking to provide a bridge between the municipal authority and traders. This innovative initiative deserves further research, but based on communication with three branch leaders and a handful of branch members, experiences with the committee (as with TUICO more generally) vary from lack of knowledge about the existence of the committee to enthusiasm – as illustrated in the quote below by an ordinary member of the Mchikichini branch. TUICO representatives acknowledge the varied receptions of the initiative amongst branches and note that it has been hard to coordinate meetings where more than a handful of representatives show up.

> We used to send our problems to local leaders for further steps to be taken because the government start with them but what we came to understand they took our claims to nowhere, so TUICO educated us, we had a committee there because there are some things the government directs but it cannot perform them, so you as business people on your own you need to come together and solve your problems otherwise if you go to market management nothing will be solved, so it needs your committee that will go and discuss the matter with the district director or anyone there.
>
> (FGD, Dar es Salaam)

The quote illustrates what could be argued to potentially be a particular strength of the trade unions, namely the ability to go – and experience with going – beyond the particular local market setting. Apart from the issue of representation, TUICO also offered collateral for some of their branches to enable them to secure bank loans and training and seminars on different issues including on entrepreneurial skills.

The approach of TUICO to enrolling members from the informal economy is to incorporate already existing groups of informal workers as union branches. KAVIMCO, for example – a group of traders selling kanga and vitenge in Mchikichini market – became incorporated as a TUICO branch, and all the members enrolled as individual union members under the KAVIMCO branch, with around 60 active members. TUICO leaves it up to the branches to decide how much informal members contribute to the union, as the traditional 2 percent salary deduction is not an option for them. Although the informal economy members do not contribute equally, all union members have equal rights in terms of voting and possibility to be elected to official posts.

Uhindin Street Machinga Association in Dodoma[10]

This association is quite illustrative of an association where members do not work in a designated trading space. It consists of around 70 micro-traders who work in the same area. There is no membership fee, but a system of contribution where members contribute TZS 2,000 when a member faces problems related to health or death in the family.

The group members previously conducted their business at the Jamatini bus stand where the association was also involved in dividing workspaces amongst the members. Some months back, they were forced to move to a new location, where they were still facing problems with the authorities as stated in the following quote:

> we have a lot of trouble from city guard who come to chase us... because this area is not for machinga so we are supposed to leave. For example last week they came to run that operation so during the whole exercise most of our properties are destructed.
>
> (KII, Dodoma)

This experience illustrates common challenges for traders who work in non-designated spaces. Even if association leaders try to defend their members, they often stand powerless when facing confiscations or forced evictions.

Association focus, rules and entry barriers

The trader associations encountered in this research appear as a continuum ranging from traders working in a similar location, to those which are larger and more robust organizations based around representation or designated trading locations to those belonging to a larger umbrella structure. Their focus varies from savings and loans over joint business activities to empowerment of women or on representing members towards local authorities.

Although associations vary, in general, limited size and financial resources means that most are vulnerable towards disruptions such as members defaulting. Depending on the accumulated capital and the size and general robustness of the association and its members, defaults might cause the association to collapse. Other challenges are relocation of members and mismanagement of funds by association leaders.

Some groups accept any member who will abide by the association rules, but most are either area or trade-specific. Some groups allowed women only, and most groups had only informal members, although a few like VIBINDO also had members with registered businesses.

Typically, the more financially oriented groups will require the ability to pay a lump-sum entry fee and the ability to save a certain amount weekly or monthly. Other groups focus more on reciprocal participation and contribution, for example, to funerals, marriages, and participation in group meetings. Finally, potential members must often be recommended by an existing member and are assessed for their financial capability and personal qualities like being trustworthy and god-fearing. Hence, it is not always an option for informal traders to become members of a well-functioning group, as noted in the quote below:

> I am currently not a member of any group because I have tried to approach some groups but as you know each and every group has not only its specific criteria for an ideal member but also specific established rules and principles so I failed to meet and sometimes to cope with some of those rules and principles.
>
> (FGD, Dodoma)

Groups have rules for excluding members who continually fail to contribute, attend meetings, or repay loans on time although the criteria for exclusion will vary between groups. Skinner (2008) when looking at informal traders' organizations has raised concerns about internal group dynamics and about representation by different segments. Although an in-depth analysis of internal dynamics within associations lies outside the scope of our research, we concur that these might not only be guided by the rules laid down in a constitution. Furthermore, the entry barriers to well-functioning groups can well make them out of reach for many traders, something that is important when considering access to social protection through associations.

Below, we discuss micro-traders' associations using the concepts elaborated in the PRA which builds onto the concepts originally conceptualized by Erik Olin Wright of structural power (derived from a worker's position in the economic system) and associational power (derived from the formation of collective organizations of workers) (Schmalz, Ludwig & Webster, 2018). The PRA was conceptualized to analyse the resources available to formal workers and starts from the basic premise that organized labour can successfully defend its interests by collectively mobilizing different power resources (Schmalz, Ludwig & Webster, 2018). Nonetheless, a broad reading of the different power resources (as discussed in more detail in Chapter 1) provides a useful tool in comparing and discussing both the actual and potential extent of different forms of power among associations of informal micro-traders.

Associational power

Based on the sheer magnitude of micro-traders and the rich variety of associational life among them, associational power seems to be potentially highly relevant. Here, the two largest associations TUICO and VIBINDO stand out with regard to their capabilities in representing member interest.

Brown et al. (2015), looking at the large evictions taking place in 2007, found VIBINDO and TUICO to have been co-opted by the establishment resulting in restricted ability to protect traders from eviction. Our research adds to this view a more varied image, where viewpoints among members range from seeing membership of the umbrella organizations as very useful to seeing it as irrelevant. Hence, our data indicate some potential in umbrella structures as VIBINDO and TUICO have been able to offer support to their affiliated groups such as access to training and representation towards the authorities. In fact, in most cases, where associations were found to have facilitated access for their members to external services (loans, training, health insurance) they were in an umbrella structure. However, in general, most associations worked on their own.

Institutional power

Within the PRA, institutional power derives from laws, regulations, procedures, and practices that regulate the relationship between worker associations and employers to facilitate social dialogue (Schmalz, Ludwig & Webster, 2018). In the case of informal traders, the relationship between associations and authorities is of key importance. Besides a few designated markets, the institutional set-up at local and national level does not include representation of informal traders except in an ad-hoc manner. Thus, in the context of micro-trade, sources of institutional power could potentially be advanced by tapping into established tripartite structures via affiliation with trade unions (at government level) and more informally tapping into the access to authorities that trade unions have at more local levels as was the case with the bargaining committee mentioned earlier. A similar potential to access the political and administrative system seemed to be in place for VIBINDO-affiliated associations, as observed by the leader of an association of coconut sellers:

> Sometimes if you have a problem or any kind of challenge with the municipal the leaders they (VIBINDO) can go there to present us ...If you face the big problem you can call the VIBINDO, or come to the office to tell them this is too big for us please support us.
>
> (KII, Dar es Salaam)

VIBINDO was also the only informal economy counterpart to NHIF in negotiating the KIKOA scheme and hence seem to have at least some access to authorities at the national level even if this is largely ad-hoc and based on personal connections.

Another form of institutional power is available to associations of members who trade in a local government-sanctioned location. Organizations which have control over their spaces and official government recognition for managing that space have legitimacy and procedures to represent their members towards market management (whether public or private). Registered associations, where members do not have a sanctioned location of trading on the other hand, are threatened by harassment and eviction. However, the power that comes with control over space is potentially time limited as a sanctioned trading place might not continue to be sanctioned under changing political leadership.

Societal power

Societal power arises from cooperation with other social groups and organizations (coalitional power) and from society's support for the demands put forward by workers (discursive power). Among micro-traders, coalitional power is seen in alliances with organizations of other types of informal workers, with trade unions, other civil society organizations, and with foreign or international NGOs and development actors. For both TUICO and VIBINDO, support from external organizations such as foreign NGOs figured prominently.

Although the informal economy, in general, and the micro-trade sector, in specific, is poorly coordinated, recently, a few attempts at broader coalitions and cooperation between different types of associations have taken place – albeit with limited success (see Riisgaard, 2020). In other countries where micro-trader associations have succeeded in influencing legislative changes towards rights and inclusion, this has most often come about either through a change in urban planning policies towards decentralization and participation (Skinner, 2008) or through extensive mobilization, coordination, and organization of micro-traders (Kumar & Singh, 2018). However, currently decentralized and inclusive planning has little political clout in Tanzania. Furthermore, the potential power that could be harnessed through the second strategy of mass-mobilization is so far left largely unexploited. A cause for some optimism is perhaps that the two largest associations TUICO and VIBINDO have signed an MOU which outlines possible areas of cooperation, although, in practice, it is yet to be implemented (KII, Dar es Salaam). This illustrates the recent tendency of trade unions proactively seeking to extend cooperation with associations of informal workers and at the same time extend their constituency. TUICO has, for example, joined StreetNet International in order to advance their representation of informal micro-traders.

Discursive power, which describes the power to successfully influence public discourses and public opinion, is not very strong in Tanzania as micro-traders are still widely seen as a nuisance and in need of formalization. However, some interesting processes can be observed relating to discursive power if understood broadly as different framings of informal traders. The favourable utterances by president Magufuli, for example, have, in a top-down manner, been shaping public discourses on micro-traders in Tanzania and might, if continued, result in increased societal power.

Another process relates to how the associations representing informal traders frame their constituency. The associations studied as part of this research ranged from viewing their members as workers – this was most prominent among the trade unions – to identifying their members as small entrepreneurs. The latter was prominent in VIBINDO which sees itself as a business association.

The framing has consequences for the kind of services and social protection that is deemed relevant for micro-traders. When framed as small businesses like VIBINDO does, emphasis is on issues like skill training, access to finance, and a conducive business environment. With regard to social insurance, emphasis is on tailor-made contributory schemes like the KIKOA. Social protection is framed largely as risk management coupled with services aimed at the business challenges of informal traders.

Another framing of micro-traders is as workers, as done by TUICO, in line with the ILO focus on decent work for all. The trade union approach in Tanzania is still uncoordinated. Thus, there is no consolidated pressure from the union federation on the government to guarantee decent work for informal workers. At the individual union level, TUICO is framing informal traders as sector-specific workers in need of representation towards the authorities.

The micro-traders commonly considered themselves as small entrepreneurs or as small business owners, while across public policies, they are commonly considered self-employed business owners. This is in line with a focus on a modern market-driven economy and frames social protection in quite a minimalist or risk management related way. Hence focus is on contributory schemes and business facilitation (via formalization) and not on employment-related rights or union representation.

Services offered by micro-traders' associations

As seen from Table 6.4, the majority (63 percent) of traders belonged to an association that they would categorize as Sacco/Vicoba/Kikundi which means that it is either related to savings and loans or support for death or health-related issues in the family.[11] About one third (30 percent) belonged to what they classify as work-related associations. As the Sacco/Vicoba/Kikundi category is not very differentiated, we also characterized association types based on what members thought was the key benefit they received from being a member. As seen in Table 6.4, this showed a similar pattern, although 49 percent mentioned loans as the main benefit, whereas 16 percent mentioned social cushioning. Voice and representation was mentioned as one of the most important benefits in Dar es Salaam only.

Roever (2014), studying street traders in 10 cities, identified three issues which membership-based organizations help with: (1) mediation with local authorities; (2) information and training; and (3) resolving conflicts among traders (Roever, 2014, p. 43). Skinner (2008), in her literature review, finds informal traders associations to focus mainly on financial services, lobbying, and advocacy. Our findings indicate that informal social insurance in case of health-related problems or death in the family should be added to the list.

Table 6.4 Key worker characteristics, association members

	Tanzania N = 166			*Dar es Salaam N = 74*		*Dodoma N = 92*	
	Sum	*Mean*	*Std. Deviation*	*Mean*	*Std. Deviation*	*Mean*	*Std. Deviation*
Formal insurance enrolment (health/ pension)	39	0.23	0.43	0.09	0.30	0.35	0.48
Association type							
Sacco/Vicoba/ Kikundi	105	0.63	0.48	0.54	0.50	0.71	0.46
Work-related	50	0.30	0.46	0.41	0.49	0.22	0.42
Women/youth/ religious	11	0.07	0.25	0.05	0.23	0.08	0.27
Benefit type							
Work-related	49	0.30	0.46	0.34	0.48	0.26	0.44
Social cushioning	26	0.16	0.37	0.14	0.34	0.17	0.38
Voice and representation	10	0.06	0.24	0.14	0.34	0.00	0.00
Loans	81	0.49	0.50	0.39	0.49	0.57	0.50
Barriers (yes = 1)	83	0.5	0.50	0.47	0.50	0.52	0.50
Association fee (yes = 1)	124	0.75	0.44	0.77	0.42	0.73	0.45

Source: Author's elaboration based on the project survey data.

Table 6.4, however, shows that about half of the associations were reported to have barriers for joining the association. While the other half were reported not to have any entry barriers, nonetheless, they might still require regular cash contributions. In fact, 75 percent of association members had to pay a membership fee to their association and 40 percent were required to pay additional regular contributions.[12]

Most of the services offered by associations can be seen as a form of informal social protection. Based on the qualitative interviews, a few particular services stand out as being very common. Most frequently mentioned was some form of informal social insurance, which almost all groups offered. This was followed by provisions of small loans, training/knowledge sharing, and representation and voice. In addition to delivering actual services, some associations also provide access to external services such as loans, training, or health insurance.

Members had most often contacted their association for help with negotiating with authorities (see Figure 6.2). This was followed by questions regarding loans, savings, or contributions and help with better terms of employment or for work-related challenges. Asked whether contacting the association helped to resolve the issue, 75 percent said yes and 19 percent answered that it partly helped resolve the issue.

In the following, we organize our discussion of the specific services offered by associations via a typology of social protection elaborated by Devereux and Sabates-Wheeler (2004). They distinguish between four main social protection types: protective, preventive, promotive, and transformative. Protective measures largely overlap with social assistance and, as mentioned, are not covered here. Preventive measures include both formal social insurance programmes' pensions, health insurance and maternity leave, and informal insurance. Promotive measures function to enhance or stabilize income, consumption, and capabilities and include access to finance or training while transformative measures

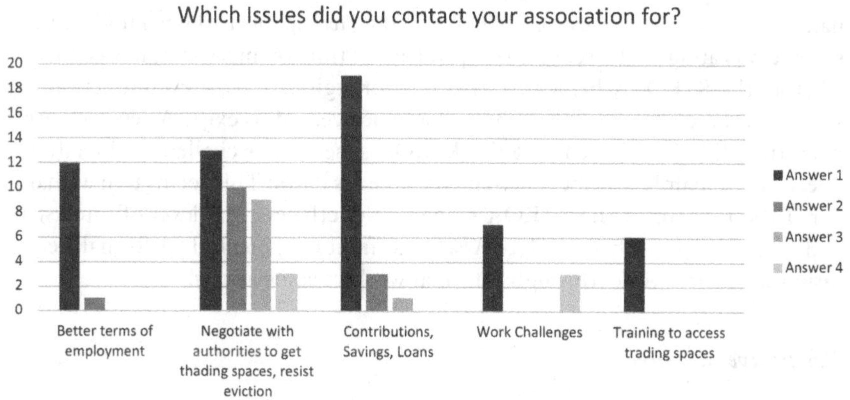

Figure 6.2 What specific issues have you mostly contacted your association about?
Source: Author's elaboration based on the project survey data.
Notes: N = 59.

address social equity and exclusion, for example, through collective action and representation.

Preventive measures

Informal social insurance

Contributions to cushion against unforeseen needs are very common in Tanzania through different forms of collective association. All participating associations offered some form of social insurance. This ranged from one-off responses to specific problems to more specific insurance schemes. One FGD participant, for example, mentioned that their association paid an amount of money for a caesarean birth but not a 'normal' one, while another group was noted to support celebratory events and minor problems. The extent of such schemes may reflect the financial capability of members as well as the size and robustness of the association. Often, the constitution would detail how much was to be received for what. One group, for example, contributed TZS 100,000 for the death of a father, mother, husband, wife, and children and TZS 50,000 for a niece, nephew, dependent, or brother-in-law (KII, Dodoma). Many groups in addition provide support by showing up for major events or by helping to host them.

No groups provided financial assistance for maternity unless it required hospitalization, but some groups gave three-month maternity leave from group responsibilities and subscription fees.

Associations facilitating access to formal social insurance

Three of the groups from Dar es Salaam had signed up to the KIKOA scheme, although not all members had subsequently enrolled. Two other groups reported to be in the process of registration, while three associations narrated how they had tried to facilitate access but failed. Several participants had been told that the scheme was about to be reorganized, and at the time of interviews, it was unclear whether the KIKOA scheme was still open for registrations or even functioning. It later became clear that the scheme has since ceased to exist. A representative from the NHIF explained that the KIKOA scheme was challenged by adverse selection, as mainly members with medical needs enrol. This seems to have hampered the roll out of the KIKOA scheme. Nonetheless, the level of interest in and engagement with the KIKOA scheme indicate potential in formal health insurance which works through informal workers associations.

Promotive measures

Loans for life contingencies or for capital

The majority of the associations offer some kind of revolving saving and loan function. Hence, 25 percent of the survey sample had received a loan within the last 12 months. Of these traders, 70 percent had gotten the loan from their associations, whereas 22 percent had received the loan from private financial institutions such as a bank.[13]

Some of the loans from associations are from the pooled monthly membership contributions and, in other cases, from stock buying. Depending on how much stock a member has bought, the member can borrow three to four times the value of the stocks at a small interest (5–10 percent). While association interest rates were considered very low, the funds available were considered insufficient to meet members' demand.

In sum, it seems that loans obtained via associations are mainly used to smoothen income and cover life contingencies. Access to finance is arguably an important component of the growth of informal micro enterprises – in our survey, 92 percent of the own-account workers expressed interest in growing their business. However, as also noted by Brown et al. (2015), access to finance is only one out of many factors which inhibits growth and thus should not be seen in isolation from other issues such as harassment, conflict over the use of public space, and other insecurities.

Training, knowledge sharing, bulk buying for members, and income generating activities organized through the associations

Members from nine associations had participated in training provided or facilitated by their associations. Training was mostly on business skills such as training on business management, on how to handle a loan or practical skills such as how to produce items for sale such as batik fabrics or soap.

> ...for example tomorrow on 29th there's a seminar. We will do it as Uwamata and will give trainings to entrepreneurs for those who will be willing. We will train on how to make handbags, peanuts and soaps making.
>
> (KII, Dar es Salaam)

Knowledge sharing would sometimes turn into joint income-generating activities as was the case with the Amani group in Dodoma which had started out as a group of women helping each other with life contingencies but which had over time developed into undertaking various joint business activities:

> So we make the batik fabrics all together and each member gets the end product as batik fabrics to go sell to other people. The member gets to keep the profit but maintains the capital for business until when we meet the following month is when we will make other items.
>
> (KII, Dodoma)

Other promotive measures included associations engaging in bulk buying for members or income generating activities such as renting out tables and pots for functions.

Transformative measures

About one out of every four associations was involved in some kind of representation of members. The scope and limitations with regard to representation have already been discussed in relation to associational power above, so suffice to emphasize here the importance of this social protection measure for micro-traders, as many of the challenges they face are related to problems with the authorities and the lack of recognition, representation, and voice.

One quarter of the associations was for women only. Although this in itself does not guarantee empowerment, sometimes, empowerment of women was an explicit aim. Quite often, helping women, in particular, was part of the reason for forming the groups in the first place. The quote below from a member of a women's only group in Dodoma illustrates how participation can lessen the dependence on husbands:

> before we use to depend on our husbands on getting money but they would just leave money for necessary domestic needs, if you get a very little amount of money that's how you're not able of reaching your goals, but now we have groups that we use to support each other, we have started our businesses, you can take-up a loan for business. So far now we don't have too many challenges because at least we get something from our groups, unlike before when we depended on our husbands.
>
> (FGD, Dodoma)

What characterizes members of associations compared to non-members?

Bromley (2000, p. 14, cited in Skinner, 2008) claims that street trader associations "typically represent older, established and licenced traders". Our results differ somewhat, in part, because the traders we sampled were very unlikely to have a business licence.[14] In order to examine the relations between association membership and key worker characteristics, we elaborate a test of means (see Table 6.5).

Table 6.5 indicates that association members compared to non-members are more likely to enrol in formal social insurance schemes, to be married, to be older, to be female, not to be wage-workers but have higher education, higher earnings, and own key assets. We note, however, that when testing these relations using a standard probit model (Riisgaard, 2020), only the variables gender, marriage status, age, and earnings remain significant. Hence, we conclude that women are more likely than men to seek out associations, that, on average, members are slightly older than non-members, and married traders are more likely to be members. It could be that the cushioning offered by associations is considered more necessary by traders with family responsibilities. The average higher earnings could point towards members being slightly better off than

Table 6.5 Differences in key workers characteristics by association member status

	Members Mean	*Non-members Mean*	*Mean Difference*	*T*
SI enrolment	0.23	0.150	0.085	1.778*
Dar es Salaam	0.45	0.50	-0.054	-0.905
Dodoma	0.55	0.50	0.054	0.905
Gender (male = 1)	0.31	0.63	-0.326	-5.769***
Married	0.69	0.39	0.301	5.297***
Local born	0.47	0.47	0.003	0.054
Age	38.02	30.46	7.566	5.976***
Daily earnings (log mean in USD)*	1.49	1.26	0.230	1.775*
Assets (house and/or land)	0.53	0.26	0.272	4.765***
Primary incomplete	0.11	0.16	-0.044	-1.077
Primary complete	0.62	0.67	-0.046	-0.801
Secondary or above	0.27	0.18	0.09	1.797*
Wage-worker	0.02	0.1	-0.076	-2.784***
Own-account	0.77	0.79	-0.021	-0.413
Micro-business	0.2	0.11	0.09	2.063**

Source: Author's elaboration based on the project survey data.

Notes: N for members = 166, N for non-members = 120. * Mean daily earnings (current USD) for members equals 8.7599 and 6.4969 for non-members. *$p < 0.10$, **$p < 0.05$, ***$p < 0.01$.

non-members although we can only speculate as to which way the causality runs. Along with the entry barriers discussed above, these differences should be taken into account when considering access to social protection through associations.

Conclusion

Public understandings of informal traders as consisting of micro-entrepreneurs, who should be assisted to graduate to formal businesses, have implications for how their access to social protection is perceived. The conceptualization of micro-traders as small businesses and not as workers is in sync with the understanding that work-related social protection does not pertain to them, even if they formalize. Micro-traders and other small entrepreneurs are instead envisioned to access social insurance such as health insurance and pensions via contributory schemes.

While enrolment in pension schemes was almost non-existent, public health insurance had become more accessible with the KIKOA scheme which worked through informal workers own associations. Despite adverse selection problems, the interest in and engagement with this scheme points to potential in health insurance schemes tailored to informal workers associations. Our findings indicated that enrolment of micro-traders in public health insurance schemes was not only inhibited by the cost but also lack of knowledge and complicated procedures. The findings furthermore indicate that formal health insurance mainly caters for the segment of informal traders who are already relatively more resilient (i.e. enrolled traders are more likely to have assets and higher education levels). In sum, existing contributory schemes are out of reach for the majority of micro-traders who have low fluctuating incomes.

The understanding of informality as a transitionary state on the way to formalization also contributes to a general lack of recognition of micro-traders as a group who should have access to representation in their own right. The perspective of people in the informal economy is sometimes addressed in an ad-hoc manner, like, for example, VIBINDO being consulted on the KIKOA scheme. However, there are very few institutionalized spaces where informal traders can directly negotiate as equal social partners, for example, in the design of social protection policies. Hence, when President Magufuli advocated allocation of trading spaces and a stop to harassment of informal traders, this is a benevolent act of the state recognizing their economic contribution and not a result of negotiations between informal traders and authorities.

Although the informal economy, in general, and the micro-trade sector, in particular, is poorly coordinated, the findings show potential in greater coordination to enhance associational power. In most cases, where associations provided representation of members or facilitated access to external services such as loans, training, or health insurance, they were in an umbrella structure. In addition, a few recent attempts at broader coalitions and cooperation between different types of associations have taken place, perhaps pointing towards a future strengthening of associational power. The increasing (although still very limited) incorporation of associations of informal traders into trade unions has to some degree opened up an institutionalized space for representation and voice for informal traders, albeit as part of the trade union movement in the established institutional model, and not as a group who should be represented in their own right. Finally, favourable statements by president Magufuli have in a top-down manner shaped public discourses towards a more favourable view on micro-traders, which could potentially be harnessed by associations to achieve increased societal power.

In general, informal trader associations in Tanzania do not have articulated policies on how informal traders should be represented in social dialogues or what a comprehensive social protection system should look like. Nonetheless, they all in their own way seek to answer to the challenges experienced by their members and provide social protection services which their members cannot access through the formal system.

It is clear from the discussion that informal trader associations offer a wider range of social protection measures than public schemes, most notably, some offer the much-needed transformative measure of voice and representation. In addition, they offer preventive measures in the form of social insurance tailored to the short-term cushioning needs of their members and promotive measures such as savings, loans, training, and joint business activities. While these measures extend only limited coverage, they nonetheless provide services which for most people in the informal economy are difficult or impossible to access elsewhere. In addition to being based on reciprocity, informal social protection among micro-traders is further characterized by being personalized, trust-based, and timely – although also exclusionary in different ways, as entry barriers to well-functioning groups can take them out of reach for more vulnerable segments of informal traders.

Overall, the findings underline the importance of taking into account the diversity of informal traders. Not only does their ability to contribute vary widely as seen by the higher occurrence of formal social insurance among traders who have assets and higher education – but so do social protection preferences which differ as women traders, for example, are more likely to prioritize membership of associations as compared to male traders. Along with the entry barriers discussed above, such differences should be taken into account when considering access to social protection through associations.

Notes

1 The majorety of the micro-traders included in the study were own-account, but wage-workers and micro-enterprise owners with one or two helpers were also included.
2 It is part of a larger research project covering informal transport, construction, and trade in Kenya and Tanzania.
3 Conversion rate = 2221.58 as of 1 December 2017.
4 See http://udadisi.blogspot.com/2018/12/is-machinga-identification-equal-to.html [Accessed 7 February 2019].
5 Unless otherwise noted, the average USD rate from June 2018 to December 2018 is used throughout the paper. The rate was 2,284.5 TZS per USD.
6 Note that 31 percent of all respondents possessed at least two assets, 13 percent possessed at least three assets, and 2 percent possessed at least four assets.
7 Highest/lowest mean = 25,127/10,852; median = 10,000/5,000.
8 Based on two interviews with VIBINDO chairman, with the chairmen of two VIBINDO member groups and with members participating in FGDs as well as secondary literature.
9 Based on interviews with two TUICO officials, a former TUICO representative, representatives from two TUICO branches and several branch members participating in FGDs.
10 Based on interview with deputy chairperson.
11 While Saccos and Vicobas are savings and loan associations registered in a specific way, Vikundi might offer some kind of savings and/or loan function or they might just offer support in case of funerals and hospitalization.
12 The maximum amount reported was 100,000 TZS (44 USD), while the mean was TZS 12,210 (5 USD) and the median TZS 3,000 (1.30 USD).
13 The remainder had received the loan from friends/relatives or from a public institution.
14 Of the own-account and micro-firms participating in the survey, 5 percent had registered their business, 1 percent had a business licence, and 1 percent had both a registration and business licence.

References

Babere, N.J. (2013) Struggle for space: Appropriation and regulation of prime locations in sustaining informal livelihoods in Dar es Salaam City, Tanzania. PhD thesis. Faculty of Humanities and Social Sciences, School of Architecture, Planning and Landscape, Newcastle University.

Bromley, R. (2000) Street Vending and Public Policy: A Global Review. *International Journal of Sociology and Social Policy*. 20 (1-2), 1–29.

Brown, A., Lyons, M. & Dankoco, I. (2009) Street traders and the emerging spaces for urban voice and citizenship in African cities. *Urban Studies.* 47 (3), 666–683.

Brown, A., Msoka, C. & Dankoco, I. (2015) A refugee in my own country: Evictions or property rights in the urban informal economy? *Urban Studies.* 52 (12), 2234–2249.

Brown, A., Mackie, P., Smith, A. & Msoka, C. (2015) *Financial inclusion and microfinance in Tanzania. Inclusive growth: Tanzania country report.* Cardiff School of Geography and Planning.

Deacon, B. (2013) *Global social policy in the making.* Bristol, Policy Press.

Devereux, S. & Sabates-Wheeler, R. (2004) *Transformative social protection.* IDS Working Paper 232. Brighton, Institute of Development Studies.

Fisher, G. (2011) Power repertoires and the transformation of Tanzanian trade unions. *Global Labour Journal.* 2 (2), 125–147.

Hickey, S. & Seekings, J. (2017) *The global politics of social protection.* WIDER Working Paper 2017/115. Helsinki, UNU-WIDER.

Hickey, S., Lavers, T., Niño-Zarazúet, M. & Seekings, J. (eds.) (2019) *The politics of social protection in Eastern and Southern Africa.* Oxford, Oxford University Press.

ILFS (2015) *Integrated Labour Force Survey 2014.* Tanzania. Available from: https://www.nbs.go.tz/tnada/index.php/catalog/31 [Accessed December 2019].

Jacob, T. & Pedersen, R.H. (2018) *Social protection in an electorally competitive environment (2): The politics of health insurance in Tanzania: ESID Working Paper 110.* Manchester, UK, The Effective States and Inclusive Development (ESID) Research Centre.

Kumar, S. & Singh, A.K. (2018) Securing, leveraging and sustaining power for street vendors in India. *Global Labour Journal.* 9 (2), 135–149.

Lyons, M. & Msoka, C. (2010) The World Bank and the street: (How) do 'Doing Business' reforms affect Tanzania's micro-traders? *Urban Studies.* 47 (5), 1079–1097.

Lyons, M., Brown, A. & Msoka, C. (2014) Do micro enterprises benefit from the 'Doing Business' reforms? The case of street-vending in Tanzania. *Urban Studies.* 51 (8), 1593–1612.

NHIF (2018) *NHIF facts and figures financial year 2017/2018.* Available from: http://nhif.or.tz/uploads/publications/en1564738202-NHIF%20Fact%20Sheet%20 2017-18-FINAL.pdf [Accessed August 2019].

Riisgaard (2020) *Worker organisation and social protection amongst informal petty traders in Tanzania.* CAE Working Paper No 4, 2020. Roskilde, Roskilde Universitet.

Roever, S. (2014) *Informal economy monitoring study sector report: Street vendors.* Cambridge, MA, WIEGO.

Schmalz, S., Ludwig, C. & Webster, E. (2018) The power resources approach: Developments and challenges. *Global Labour Journal.* 9 (2), 113–134.

Skinner, C. (2008) Street trade in Africa: A review. Working Paper No 5. WIEGO.

Tripp, A.M. (2000) Political reform in Tanzania: The struggle for associational autonomy. *Comparative Politics.* 32 (2), 191–214.

Tsuruta, T. (2006) African imaginations of moral economy: Notes on indigenous economic concepts and practices in Tanzania. *African Studies Quarterly.* 9 (1–2), 103–121.

7 Access to social protection in Kenya

The role of micro-traders' associations

Raphael Indimuli

Introduction

In Sub-Saharan Africa, micro-trading[1] is the most visible form of occupation (Brown, 2006), accounting for an estimated 43 percent of all informal non-agricultural employment (Skinner, 2008). It is a source of livelihoods dominated by women traders (Chen, Roever & Skinner, 2016). In Kenya, the sector falls under the Micro, Small, and Medium Enterprises (MSMEs), a sector comprising of licensed and unlicensed businesses. The vast majority (93.3 percent) of the unlicensed businesses are micro, engaging less than ten employees and having an annual turnover of less than Kenya Shilling (KES) 50,000 or an equivalent of US dollar (USD) 494.1.[2] Of the 18 sub-sectors within the MSME sector, wholesale and retail trade is the largest sub-sector (62.9 percent) due to low entry barriers. The sub-sector accounts for more than half (53 percent) of informal employment and contributes about 22.8 percent to Kenya's Gross Domestic Product (KNBS, 2016).

Despite the contribution of IWs to Kenya's economy and to employment creation across much of SSA, research on social protection reveals that most IWs in Africa lack social protection (ILO, 2017). The micro-traders who are the focus of this chapter work under deplorable conditions (Mitullah, 2003) without sanitation, social protection such as sick leave, paid leave, and maternity benefits. They work long hours, and their incomes are low and irregular. Their environment makes them vulnerable to ill health while inadequate incomes limit their ability to accumulate savings (and invest in assets) and to take up formal social protection products such as health insurance and pensions (Skinner, 2008; Chen, Roever & Skinner, 2016).

Recent research also reveals that micro-traders in developing cities are subjected to endless harassment from city officials (Chen, Roever & Skinner, 2016). Scholars (Lindell, 2010) point out that several IW associations have emerged to articulate their grievances and claim rights to work in public spaces. However, the literature on IWs and on social protection gives little attention on the possible link between social protection and IWs associations. Social protection include all measures, formal and informal that seek to provide relief from deprivation such as resource transfers in cash or in kind (protective measures); measures that seek to avert deprivation such as health insurance and pensions (preventive measures);

DOI: 10.4324/9781003173694-7

measures that seek to improve real incomes and enhance capabilities such as credit schemes (promotive measures); and measures that seek to address issues of social equity and exclusion such as collective action (transformative measures) (Devereux & Sabates-Wheeler, 2004). This chapter adopts Devereux and Sabates Wheelers (2004) framework to generate new knowledge on informal trader associations and their potential for enabling access to both formal and informal social protection measures in Kenya. However, it does not address protective measures since most IW associations do not offer this form of social protection.

This chapter uses empirical data collected from a combination of quantitative and qualitative methods, notably: a survey, Focus Group Discussions (FGD) and Key Informant Interviews (KIIs). It draws data from a survey that was conducted between April 2018 and December 2018 which interviewed 231 micro-traders in Nairobi and Kisumu. Three quarters (73.5 percent) of the survey interviews were sampled geographically, while the remaining quarter (26.5 percent) was sampled through a variety of traders' associations. The survey covered vulnerable groups of micro-traders, including traders who do not have fixed locations (mobile traders) and those traders without permanent or fixed structures such as stalls. Three types of micro-traders were identified: paid workers, own-account workers, and micro-enterprise owners with a maximum of two employees. The KIIs and FGDs were conducted between October 2018 and December 2019. A total of five FGDs were conducted with members of associations, while a total of 11 KIIs were conducted with leaders of those associations.

All economic activities in Kenya, including those of IWs, are governed by the state. The Kenyan Constitution provides for two levels of governance: the national level and the county level. At the national level, national policies and legislation recognize the contribution of IWs to the economy. This recognition dates back to 1972 when the International Labour Organization (ILO) conducted a study on employment and incomes in Kenya (ILO, 1972). Since 1972 to date, several policies have been introduced to address challenges associated with informality. These policies use terms like 'jua kali'[3] sector, informal sector, small-scale sector, and SME sector to denote 'informality'. Sessional Paper No. 2 of 1992 on "Small Enterprise and Jua-kali Development" was a pioneer policy exclusive to the sector. The policy defined "small enterprise" as an enterprise employing between 1 and 50 employees (ROK, 1992). Later, the Sessional Paper No. 2 of 2005 on "Development of Micro and Small enterprises for wealth and employment creation" was developed to address issues that prevent micro and small businesses from formalizing – such as registration, taxation, and licensing (ROK, 2005).

The MSE Act of 2012 is the first comprehensive law that is specific on informal actors. It provides for MSEs and establishes structures like the Micro and Small Enterprise Authority (MSEA) whose main function is to govern informal actors as well as the MSE Fund aimed at offering affordable finance to SMEs. Informal workers are represented on the board of MSEA, and its structures[4] provide a platform for IWs to participate in the making of decisions that affect their work sites. Furthermore, a draft bill on Street Vendors Protection of Livelihood Bill (ROK,

2019) was at the Senate with the object of granting rights to traders to engage in micro-trading in designated areas without harassment from city officials. It proposes the creation of a Hawkers and Street Vendors Authority (HSVA) whose task will be to manage the affairs of hawkers and street traders in Kenya.

Governance of micro-trading falls directly under city authorities (Skinner, 2008). The cities of Nairobi and Kisumu covered in this analysis are major cities in Kenya. Nairobi is the capital and largest city, while Kisumu, which is an inland port city, is the third largest urban centre. Cities, under the Urban Areas and Cities (UAC) Act of 2011, are to be governed and managed by County Governments with the help of Urban Boards (NCLR, 2019). County Governments require all micro-traders under their jurisdiction to apply for licenses or permits (KNBS, 2016). Application for licenses requires micro-traders to indicate trade location and indicate physical address, which is not possible for mobile traders. City authorities also expect traders to renew their license every year. Yet, most traders operate without it (KNBS, 2016). Some scholars (Racaud, Kago & Owuor, 2018) argue that the policy environment in Kenya discourages traders from acquiring licenses because acquisition of a license does not guarantee that city authorities will not evict them. Racaud, Kago, and Owuor further argue that ambiguity in policy and practices leads to conflict between city authorities and traders. However, in the past few years, County Governments of Nairobi and Kisumu have shown more tolerance for micro-traders' activities. For instance, in Nairobi, the County government permits micro-traders to transact business without licenses in the Central Business District (CBD) on specific days or times (i.e. evenings) or at designated locations. The more tolerant atmosphere in Nairobi stems from the public statement issued by County Governor of Nairobi in 2017 abolishing collection of fees from 'mama mbogas' (Swahili term for female vegetable vendors) (DN, 2017). The Governor argued that abolishing the fee was a fulfilment of one of his campaign promises. In Kisumu, micro-traders are permitted to transact businesses on designated streets within the CBD, and most pay a trading fee to city officials as discussed later.

How informal workers are treated can be linked to perceptions of informal work by stakeholders, including the state. While there is no universally accepted definition or description of informality, the ILO, a global body that promotes the decent work agenda which includes right to social protection, describes informality as all economic activities by workers and economic units that are not covered, or not adequately covered, by formal arrangements in law or in practice (ILO, 2002). In Kenya, informality tends to be understood from an enterprise point of view, as unlicensed or unregistered enterprises (KNBS, 2016). Informality in policy documents is referred to by terms such as informal sector, *jua kali* sector, and the SME sector (ROK, 1992, 2005). Amidst these interpretations, it was necessary to find out from micro-traders how they perceive themselves. Discussions with micro-traders revealed that most perceived themselves as informal using various phrases like "we are '*jua kali*'", "we are harassed by city officials", "we do not have a salary", "we do not have contracts", "we lack formal benefits e.g. health insurance and pensions", among others. Evidently, these quotes illustrate that the view of informality relates to location of operation, taxation, registration, condition of work, and the lack of social and legal protection. While the term 'workers' is used, it is important to note that most micro-traders did not perceive themselves as workers *per se* but as entrepreneurs or businesspersons. This view is also reflected in national policies that interpret informal traders as the self-employed and as people operating or employed in unlicensed or unregistered enterprises (KNBS, 2016).

The remainder of this chapter is organized in the following way: it commences by a section on 'socio-economic characteristics' which provides a detailed description of Kenyan micro-traders by focusing on demographic, social and economic features, and discusses key features of their work. It is followed by a section on 'micro-trader challenges' which briefly outlines the challenges in micro-trade to establish the need for social protection, and a section outlining the role and challenges of trader associations and provides a link between trader associations and formal insurance. Thereafter, a discussion section follows which compares members of associations with non-members and micro-traders enrolled in formal insurance schemes with the non-enrolled. Last, concluding remarks are made in conclusion section.

Socio-economic characteristics of micro-traders

Demographic characteristics have an influence on the uptake of formal insurance and/or enrolment in associations. Table 7.1 presents a set of worker characteristics relevant for understanding such relations. The full sample count is 223 workers, out of which 57 percent were sampled in Nairobi and 43 percent from Kisumu.

Demographic characteristics

Table 7.1 shows that almost half of micro-traders belong to an association, although some belong to more than one. Region-wise, more than half of sampled traders in Kisumu are members of associations compared to 43 percent in Nairobi. In terms of formal health insurance coverage, 39 percent of micro-traders have either their own cover or through their spouse. Interestingly, this is higher than the 2015/2016 Kenya Integrated Household Budget Survey (KIHBS), where 19 percent of the population, at the national level, had health insurance (KNBS, 2018). The dominant health insurance (93.3 percent) identified in the KIHBS report (also in our study as we shall see later) is the National Health Insurance Fund (NHIF) cover. The KIHBS report reveals urban populations as having a higher share of health insurance (29.2 percent) than rural populations (13.3 percent): the share of population with health insurance cover is even higher for Nairobi (40.7 percent) compared to Kisumu (27 percent). This finding seems to be in line with our finding on Nairobi (41 percent), but, for Kisumu, our findings are much lower (36 percent).

Table 7.1 Key worker characteristics

	All		*Nairobi*		*Kisumu*	
	Mean	*SD*	*Mean*	*SD*	*Mean*	*SD*
Association membership*	0.47	0.50	0.43	0.50	0.53	0.50
Formal SI (health/pension) enrolment	0.27	0.45	0.26	0.44	0.29	0.46
Heath insurance coverage	0.39	0.49	0.41	0.49	0.36	0.48
Gender (male = 1)	0.45	0.50	0.46	0.50	0.43	0.50
Age	36.65	10.92	36.33	11.76	37.05	9.76
Married	0.63	0.48	0.56	0.50	0.72	0.45
Local born	0.28	0.45	0.18	0.39	0.41	0.49
Mean daily earnings** (USD)	11.12	11.61	8.70	7.05	14.27	15.14
Assets (house and/or land)	0.23	0.42	0.17	0.37	0.32	0.47
Primary incomplete	0.16	0.37	0.12	0.33	0.22	0.41
Primary complete	0.40	0.49	0.44	0.50	0.35	0.48
Secondary and above	0.43	0.50	0.44	0.50	0.43	0.50
Training course	0.04	0.21	0.06	0.23	0.03	0.17
Training on-job	0.20	0.40	0.25	0.43	0.14	0.35
Training self-taught	0.75	0.43	0.70	0.46	0.82	0.38
Wageworker	0.05	0.23	0.07	0.26	0.03	0.17
Own-account	0.81	0.40	0.81	0.39	0.80	0.40
Micro-enterprise	0.14	0.35	0.12	0.33	0.16	0.37
Observations	223		126		97	

Source: Author's elaboration based on project survey data.
Notes: * figures are from the random sample where N = 164, n (Nairobi) is 96 and n (Kisumu) is 68
** median daily earning is USD 7.4 which is often a more accurate measure of income than mean when SD is high. Median daily earning is USD 6.7 for Nairobi and USD 7.9 for Kisumu.

Table 7.1 also reveals that only 27 percent are making individual contributions to formal social insurance schemes, that is, health insurance and pension, slightly higher in Kisumu (29 percent) compared to Nairobi (26 percent). The differences in contribution patterns are partly tied to incomes as discussed later. Women constitute 55 percent of the micro-traders, which is consistent with national results showing that a large share (60.7 percent) of unlicensed enterprises in wholesale and retail are owned by females (KNBS, 2016).

The average age of micro-traders is 37, most of them are married (63 percent), and the majority have obtained formal education with 43 percent having completed secondary education. These findings are consistent with the national findings showing that a higher share of business operators (59.7 percent) under wholesale and retail had completed secondary education, followed by 30.7 percent who had primary education, and 9.6 percent who had no education (KNBS, 2016). With respect to skills, a large segment of micro-traders (75 percent) are self-taught, 20 percent learnt the skills while on the job, and the remaining 5 percent attended a training course. These findings illustrate that micro-trading requires no specific vocational training or only a brief initiation.

Income and assets

Table 7.1 shows that the average daily earnings of micro-traders is USD 11; however, the median value (USD 7.4) offers a fairer representation due to some high earners in the sample pulling up the mean. In terms of region, Kisumu traders have higher incomes than Nairobi traders, and this difference is statistically significant. Likely explanations for this difference are: (i) some food products sold in Kisumu such as fish are likely to fetch high returns, (ii) as mentioned earlier, Kisumu county government permits traders to transact businesses in designated commercial streets, including moving where there is human traffic (near bus terminus) after 5 pm and this is likely to improve profits, and (iii) the Kisumu sample composed of fewer wage-workers who have lower mean earnings (USD 3.36) compared to micro-enterprise owners (USD 19.89). This difference in income may partly explain why the proportion of those making contributions to formal insurance schemes was higher in Kisumu (29 percent) than in Nairobi (26 percent).

In terms of assets, Table 7.1 reveals that 23 percent of workers owned a plot of land and/or a house. Region-wise, a higher share of Kisumu traders (32 percent) own assets compared to Nairobi traders (17 percent), and this difference is statistically significant at the 99-percent confidence level. This is in accordance with Kisumu traders having higher incomes, and since assets may be used as collateral for loan applications, that explains the observed higher share of Kisumu traders (40 percent) applying for loans compared to Nairobi traders (33 percent).

Work-related characteristics

Table 7.1 reveals that most micro-traders (81 percent) are own-account workers, 14 percent are micro-enterprise owners, and only 5 percent are wage-workers. Since access to formal insurance benefits is tied to an employment relationship under ILO Convention No. 102 (WIEGO, 2019), our findings suggest the need to rethink the design of employment-related protection since most traders are not in a standard employment relationship. Most traders operate informally since they are neither registered nor licensed. The rest of businesses are partly informal, that is, registered only or licensed, supporting the view of informality as a continuum and not a dualistic sector.

While Kenya seeks to formalize the informal economy through registration and licensing, our findings show that an overwhelming majority of the micro-traders (97 percent) have not applied for business registration. Since the law prohibits one from operating without a license, many pay a fee ranging from between KES 30 and KES 50 to urban authority officials to occupy their trading locations (54 percent), to prevent harassment (33 percent) or eviction (9 percent) from those locations.

In terms of region, most Kisumu traders pay official fees compared to 10 percent of Nairobi traders as shown in Figure 7.1. Thus, the activities of traders are more regulated in Kisumu, whereas most of the Nairobi traders (60 percent)

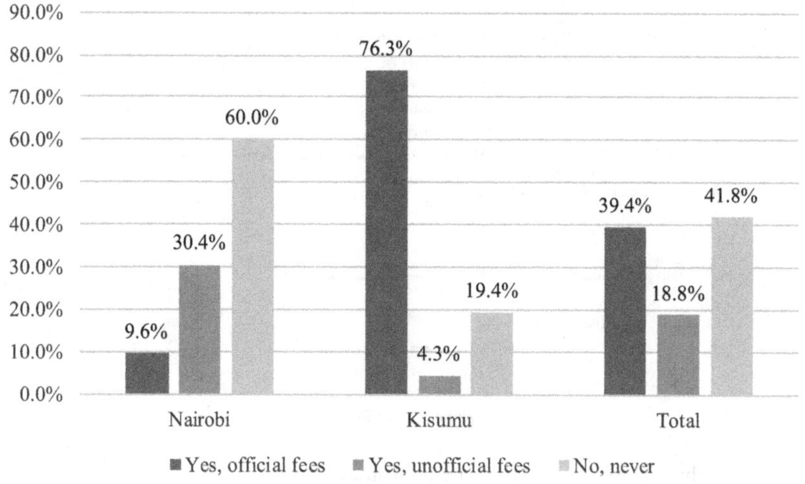

Figure 7.1 Distribution on payment of fee by city.
Source: Author's elaboration based on project survey data.

do not pay any fees which can perhaps be partly explained by the Nairobi City County Government's decision to abolish fees charged on small-scale traders as was noted earlier.

Reasons for engagement in micro-trade

Most micro-traders (73 percent) operated close to their homes at a distance of under 5 km. Analysis of distance covered by gender revealed that a higher share of women (79 percent) lived closer to their homes than men (65 percent). This indicates that women are more likely to take up jobs closer to their homes due to their reproductive roles, which also explains why a larger share of micro-traders are women.

Table 7.1 shows that Kisumu has a higher share of micro-traders who are locally born (41 percent) compared to Nairobi with 18 percent indicating that a large share of micro-traders migrated to the capital in search of employment opportunities. In contrast, since Kisumu city is a smaller city (with fewer employment opportunities compared to Nairobi), it is likely that those traders that were not born in Kisumu city migrated from rural parts of Kisumu county or nearby counties.

Dualists (Hart, 1973) argue that people migrating to urban areas but do not find formal work are likely to find employment in the informal economy. Our finding partly confirms this claim as noted by many micro-traders. Figure 7.2 shows that the two main reasons were "money is good" and "I could not find anything to do". On the second response, the top two reasons were "I prefer to be my own boss" and "I could not find anything to do". The prevalence of

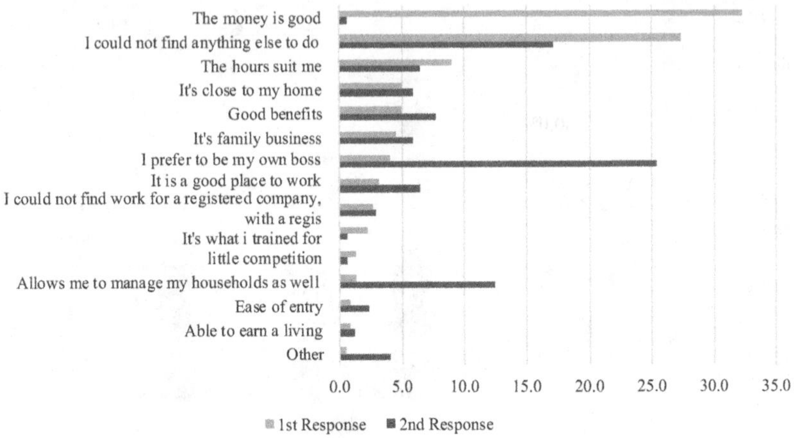

Figure 7.2 Why did you choose this line of business?
Source: Author's elaboration based on project survey data.

"I could not find anything to do" is evidence that people are getting into micro-trading due to lack of employment opportunities in the formal economy. The reasons "money is good" and "I prefer to be my own boss" lend support to the voluntarist school of thought (e.g. Maloney, 2004) who argue that some people willingly choose to work in the informal economy because they perceive opportunities such as higher incomes but not because they are forced by circumstances. The share of those who indicated that 'money is good' was higher among micro-enterprise owners (48 percent) than own-account workers (31 percent) and paid workers (17 percent). This finding suggests that job satisfaction is positively correlated with earnings since micro-enterprise owners tend to have higher incomes compared to own-account workers and paid workers. The next section outlines the challenges faced by micro-traders to establish the need for social protection.

Micro-traders' challenges

Micro-traders face many challenges, which include harassment by city officials (24 percent), lack of shelter (22 percent), inadequate capital or high prices of raw materials (14 percent), reduced customers or sales (11 percent), among others as shown in Table 7.2. On the second challenge, problems tied to lack of shelter topped the list as shown in second column in Table 7.2. The harassment complaint confirms the conflict between traders and city authorities highlighting the need for voice and representation. The challenges of heat and weather conditions are tied to the nature of trading locations and legislative requirements. Micro-traders are prohibited from making permanent shelters forcing some to operate from open spaces and others to move with their wares from place to

Table 7.2 Describe the two key challenges/threats/problems you face in your work

	Challenge 1 (Percent)	*Challenge 2 (Percent)*
Reduced customers/low sales	11.2	9.4
Harassment from city officials	23.8	3.6
customer debts	2.7	1.8
High prices of goods/lack of raw materials/inadequate capital	13.9	8.5
Competition from other vendors	5.4	3.1
Harsh weather, for example, rain, cold/no shelter	22.0	14.3
Losses due to perishable/seasonal products	10.8	4.9
Forced to close shop due to parental obligation/ sickness	1.3	–
Lack of storage facilities/no business location	1.8	4.0
No challenge	4.9	47.1
Insecurity/theft	0.4	1.8
Air pollution	0.9	0.9
Waste disposal	0.9	–
Lack/inadequate clean water	–	0.9
Total (N = 223)	100.0	100.0

Source: Author's elaboration based on project survey data.

place. Those located along roads are predisposed to vehicle emissions which negatively impact on their health. These findings reveal that micro-traders are operating in harsh conditions, and their challenges require policy attention and support in terms of access to social protection.

The goal of OHS is to foster a safe work environment. However, this framework has traditionally been applied to formal employment, with marginal attention to IWs. Most micro-traders (76 percent) have never been injured at their workplace or suffered from a work-related illness, while 24 percent reported having suffered from work-related injury or sickness, and of those, the majority (65 percent) were from the Nairobi region. This suggests that micro-trading carries little risk of personal injury from use of tools at work. However, some of their trade locations are prone to harassment, crime, and air pollution. Harassment which emanates from the conflict between traders and city authority officials is often violent, especially within the CBD (Racaud, Kago & Owuor, 2018).

Our results show that most of those who suffered from work-related injury/sickness incurred medical expenses (42 out of 54) and of those the majority paid from out of pocket (28 out of 54). The rest were paid for by relatives, employer, and colleagues. Only a few of those who reported having suffered from work-related injury/sickness used health insurance cover.

Slightly above half (28 of 54) continued working despite injury or work-related illness. The rest stayed away from work for less than a month and only a few for more than a month. This finding suggests that micro-trading does not present serious OHS risks forcing micro-traders to abandon work, while others are forced to work because of lack of options.

Micro-trader associations

This section focuses on associations formed by micro-traders with a view to showing if they enable access to formal and informal social protection measures.

Association landscape

Although there are many types of associations, the dominant associations among micro-traders are *chama*[5] (67 percent), worker associations (22 percent), and others (11 percent) as shown in Table 7.3. In terms of region, Nairobi has a higher share of *chamas* (79 percent) compared to Kisumu (54 percent), and a significant share of worker associations are in Kisumu (31 percent) compared to Nairobi (14 percent).

Table 7.3 reveals that the share of association members having formal health insurance is higher (49 percent) compared to the sample average (39 percent in Table 7.1). It also reveals a higher share of association members as contributors to formal insurance schemes (35 percent) compared to the sample mean of 29 percent indicating a link between belonging to associations and access to formal insurance. In terms of region, health insurance coverage is higher in Nairobi (54 percent) than in Kisumu (45 percent), and this is tied to the fact that 16.7 percent of Nairobi traders indicated that they were under their spouse's cover compared to only 9.3 percent Kisumu traders.

Table 7.3 Key worker characteristics, association members

	All		*Nairobi*		*Kisumu*	
	Mean	*SD*	*Mean*	*SD*	*Mean*	*SD*
Formal insurance enrolment	0.35	0.48	0.31	0.47	0.40	0.49
Health Insurance coverage	0.49	0.50	0.54	0.50	0.45	0.50
Association type						
Sacco/vicoba/chama	0.67	0.47	0.79	0.41	0.54	0.50
Worker association	0.22	0.42	0.14	0.35	0.31	0.47
Other association	0.11	0.31	0.07	0.26	0.15	0.36
Benefit type						
Work-related	0.15	0.36	0.14	0.35	0.15	0.36
Social cushioning	0.01	0.09	0.00	0.00	0.02	0.12
Voice	0.01	0.12	0.00	0.00	0.03	0.17
Loans	0.83	0.38	0.86	0.35	0.80	0.40
Barriers	0.65	0.48	0.65	0.48	0.65	0.48
Fee	0.74	0.44	0.63	0.49	0.86	0.35
Observations	136		71		65	

Source: Author's elaboration based on project survey data.

Trader associations in Kenya can be categorized into two: grassroot level associations and umbrella associations which function either at the national or county level. Grassroot associations are quite small comprising at least 10 members and mainly address issues relating to welfare of members such as sickness, funerals, school fee payment, and purchase of household goods and offer opportunities for savings and access to loans. On the other hand, umbrella associations have large membership comprising a network of grassroot associations, whose main role is advocacy and representation.

The next section provides brief histories of three cases of associations. The first case is an example of a national umbrella association playing a unique role of voice and representation, while the second is an example of an association focusing on asset creation as the basis for organizing. The last case is an example of a grassroot association.

Association histories

The Kenya National Alliance of Street Vendors and Informal Traders (KENASVIT)[6] is an umbrella association formed in 2005 to represent and voice concerns of street vendors, hawkers, and market traders in Kenya. It brings together 20 urban affiliates from across the country. Its activities are guided by a constitution which outlines leadership structure and objects of the association. Its leaders are drawn from the chapters and elected for a term of four years.

The KENASVIT Chairperson sits as a member of MSEA as illustrated in the following quote: "...my role in the board is to advise the government on issues pertaining to MSE and to 'encourage Kenyans to be entrepreneurs'" (KII, Nairobi).

KENASVIT has about 400,000 mostly female members. Registration for groups is KES 5,000, and individual members pay monthly subscription fees of KES 20 through their groups. KENASVIT advocates for rights of traders to use public space. This role is more pronounced due to regular conflicts between traders and city authorities. In 2014, for example, when traders in Kisumu were evicted from Oile market, KENASVIT leaders engaged Kisumu County officials, but their efforts did not prevent the eviction. KENSAVIT Chair noted that there are many competing associations of traders preventing them from having a unified voice. Traders tend to be more united during crises such as evictions. In the absence of conflict, support of umbrella associations declines as expressed in this quote: "...when members are being harassed, that is when they look for me. They will call you at any time when their businesses are being threatened. However, when things go back to normal, they relax".

While protest is a common approach to defending rights, KENASVIT employs other strategies such as negotiation, as observed in this quote, "...negotiation allows for give and take. You do not stand your ground". Although traders, through this approach, have obtained rights to use public spaces, trading locations are not allocated to them permanently but temporary as they might be required to vacate if need be.

Through networking with other trader associations, KENASVIT has been able to prevent eviction of traders in other areas of Kisumu. Some of KENAVIT partners are local grassroot organizations (e.g. Muungano wa Wana vijiji), national NGOs (Pamoja Trust and Kenya Land Alliance), public agencies, for example, the Institute for Development Studies (IDS), MSEA, and transnational organizations, for example, StreetNet International, Women in Informal Employment Globalizing and Organizing (WIEGO), and Unitarian Universalist Service Committee (UUSC), and other international development partners (e.g. DFID and Oxfam). These partners play a crucial role providing financial and technical support.

Micro and Small Enterprise Leaders' Summit (MSEL)[7] was registered in 2013 and has over 60,000 members but only 30,000 are active. Most active members were noted to be female (20,000). Half of MSEL members are informal traders. To become a member, a group must pay a registration fee of KES 3,000, while an individual must pay a fee of KES 1,000. The association has members in all the 17 constituencies of Nairobi City County and holds election yearly. The term of elected officials is renewable every year. However, there is no limit to the number of years an official can serve in MSEL.

Prior to registration, MSEL existed as a movement. Currently, it is an umbrella association of groups and Community Based Organizations (CBOs). Membership is open not only to groups and CBOs but also to individuals. MSEL not only performs the role of voice and representation but also empowers members economically and socially. Members comprise self-employed workers. In terms of gender, most of the MSEL members are women. Age-wise, MSEL is dominated by youth.

MSEL operates Savings and Credit Cooperatives (SACCOs)[8] in each constituency of Nairobi. To become a SACCO member, one must register with KES 500. However, membership to the SACCOs is voluntary. According to MSEL Chairperson, the purpose of establishing SACCOs was to encourage members to save regularly and own assets, hence be able to access loans from financial institutions. Members of SACCOs contribute weekly as subscription fee to service loans. The association had saved up to KES 30 million, which enabled it to access a loan of KES 25 million to buy land for building houses for SACCO members.

MSEL is working with other institutions at the local, national, and international level. At the international level, MSEL in collaboration with county government participated in one of the United Nation (UN) meetings on green economy to promote tree planting. At the national level, cage fish farmers participated in the blue economy conference which took place early 2019.

MSEL also advocates for the rights of traders. It advocated for mobile traders to trade in Eastleigh after the business community of Eastleigh had opposed trade in front of their premises. This resulted in a policy providing space on the third floor of a commercial building for mobile traders. In addition, MSEL spearheaded the establishment of Eastleigh Hawkers Association (EHA).

Apindi Smart Friends (ASF)[9] is a grassroot association based in Kisumu, established in 2013, and is registered with the Ministry of Gender, Children and

Social Development which registered *chamas*. It comprises 30 members but only half are active and committed to making regular contributions to the group. In terms of gender distribution, eight of the active members are males, while seven are females. The group is open to all, as expressed by the Chairperson, "... anyone can join. We have street traders, teachers, transport workers, family and friends joining the group. But majority of members operate from the street and contribute daily fees to the County government". The group mobilizes resources through savings and issues loans to members at a small interest rate. It also has an emergency fund which members contribute KES 50 daily and an income generating activity involving hiring of chairs to other groups for meetings. The group cushions members in case a member falls sick and is admitted to hospital through contributions. Each member is required to contribute KES 300, and very few members of the group have access to formal social protection services.

Social protection services

This section uses Devereux and Sabates-Wheeler (2004) classification of social protection measures as an analytical framework for understanding services offered by associations and formal insurance schemes.

Preventive measures

According to Devereux and Sabates-Wheeler, preventive social protection measures are measures which seek to avert deprivation. Examples of such measures include health insurance and pension. Services offered by the state or by associations that seek to avert financial stress attributed to eventualities such as sickness, death, or loss of employment fall within this category.

Access to formal social insurance:

The two main avenues for accessing health insurance and pension services offered by the state are through the NHIF and the NSSF schemes. Participation of IWs in these schemes is voluntary, whereas, for formal workers, it is mandatory. In the absence of an employer, IWs must self-enrol and make monthly contributions to the schemes to access services. As earlier mentioned, only 27 percent contributed regularly to a formal scheme and of these most contributed to a health insurance scheme (50 out of 61), but only a few contributed to a pension scheme (6 out of 61) and other insurance schemes (5 out of 61); a clear indication that health is a priority for traders rather than saving for retirement.

Figure 7.3 shows that the two top causes for low enrolment in formal insurance schemes are cost and lack of interest. This finding suggest that most micro-traders find it difficult to pay monthly premiums because of irregular, low incomes and the premium cost which is considered too high.

A probe of poor enrolment in formal insurance schemes with association members revealed that, for the NSSF scheme, inadequate, or lack of, information is the main reason for poor enrolment. Most participants mistook the NSSF scheme to be a product for formal workers. Others noted that they had lost trust

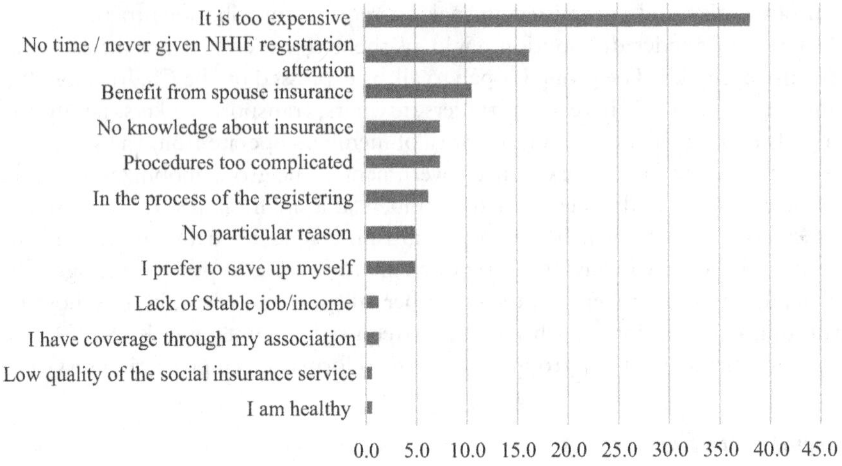

Figure 7.3 Reasons for not enrolling.
Source: Author's elaboration based on project survey data.

in the NSSF scheme due to corruption. One informant indicated that pension was not a priority for them as quoted, "…we are businesspeople. We find it difficult to save for retirement when you still have stock to buy" (KII, Nairobi). This quote shows the pension schemes as unattractive and unsuited to micro-traders' needs.

Regarding the NHIF, some indicated that the monthly premium of KES 500 was too high, while others found the registration processes as cumbersome because it requires them to take time off from work to register which means loss of income. The perception of insurance as a 'bad omen' was also an obstacle to enrolment as is noted in the following quote:

> …some people do not join because they believe insurance comes with a bad omen. There is Deputy Commissioner who is my friend who believes that having an insurance brings bad omen. I know a number of people who share the same belief.
>
> (KII, Nairobi)

Almost all micro-traders (95 percent) knew about the NHIF scheme but not necessarily how it works. Interestingly, almost half (48 percent) came to know about the scheme from their social networks such as family, friends, and from their associations. This shows that social networks are a crucial channel in the spread of information. The rest knew about the NHIF from media (29 percent), interaction with NHIF officials (10 percent), or interaction with hospital staff (5 percent) or from previous employment (8 percent).

Of the 39 percent in Table 7.1 that are members of the NHIF scheme, many (42 out of 87) registered themselves, some (30 out of 87) registered through their spouse, and others (15 out of 87) registered themselves and their spouses. This finding shows that enrolment is not only an individual decision but also a social decision.

Among the non-members of NHIF scheme, a few (25 out of 136) had previously been members of the scheme. When these were asked to give reasons for leaving the scheme, their reasons were 'it was too expensive' (60 percent), 'loss of formal employment' (20 percent), disappointed with the scheme benefit (12 percent), benefit from spouse (4 percent), and no reason (4 percent). During discussion with micro-traders and KIIs, this issue was probed. Box 7.1 confirms the above reasons:

The above quotes reveal that default decisions are influenced by factors related to socio-economic characteristics such as low and irregular incomes and factors related to service provision such as cost of premium and cumbersome procedures. This suggests the need to redesign the NHIF scheme to make it attractive

Box 7.1 Reasons for default

...I had an accident and could not pay for NHIF. I paid the fine. The payment of the fine of KES 250 is a problem (FGD, Kisumu).

...sometimes we wonder, what is the use of having the card? I am forced to pay for medicine that is worth more than KES 700, yet I have a card (FGD Nairobi).

...the deadline for remitting the money is 9th of every month...try paying (via m-pesa) on 7th or 8th, and it will not reflect in the system (FGD, Nairobi).

...there is no client service in public hospitals. They mistreat us. Nurses are very bad. The services are poor. The machines are sometimes not working. However, the private hospitals (where we pay cash) treat us very well (FGD, Nairobi).

...I work in town, but I have registered to access hospital where I live. If something happens to me in town, I must go to where I registered to receive services. If I am sick in the rural areas, I must travel to Nairobi to receive services (KII, Nairobi).

...it is difficult for NHIF members to change hospitals. Changes can only be made on the months of June and December (FGD, Nairobi).

...if you are sick and you want approval from NHIF, the process is hard. They take you to and fro. For MRI clearance you are sent to headquarters (at Upperhill) to get approval (FGD, Nairobi).

in terms of cost and to adjust the premium payment plan to the realities of informal work characterized by low and irregular incomes and long working hours.

Informal social protection measures:

Discussions with association members revealed that most associations had set up structures (e.g. rules and social funds) to cushion members from risks associated to sickness, death, and loss of employment as is noted:

> …we contribute towards welfare. If you are sick or have any problem such as funeral, lacking school fees. The group discusses and agrees to give you some money.
>
> (FGD, Kisumu)

Members mobilized resources in forms of savings on a daily, weekly, or monthly basis. This practice was anchored in a group constitution and members who defied it were subject to some form of sanctions. Almost half of association members (45 percent) in the survey contacted their associations to access the services offered by the associations, mostly services related to welfare and benefits (92 percent). Some groups offered some form of 'maternity leave' to nursing mothers as noted by this quote, "…new mothers are given 2 months maternity leave which entails not attending meetings" (FGD, Kisumu). Other groups assist women in managing their businesses while on maternity to ensure that they have an income and are consistent in payment of contributions to the group. Some other groups of women contribute a given amount of money to take care of expenses of the nursing mothers.

Although groups cushioned members during sickness, it was noted that not all kinds of illness were covered by groups but only those that lead to hospital admission or those that prevent members from working. Some groups could visit sick member at home or in hospital while others contribute to offset hospital bills on admission of their member. However, the assistance given to members was noted to be one-off hence most groups had no provision for prolonged hospital admission.

Promotive measures

Promotive social protection measures are those measures which seek to enhance incomes (Devereux & Sabates-Wheeler, 2004). One can argue that services offered by groups such as loans and training fall under promotive measures. Most of those belonging to associations (83 percent) reported savings and loans as the most important benefit of being in an association as shown in Table 7.3. The ability to issue emergency loans and flexible payment plans are some of the features of associations as is noted:

> Recently a member lost goods and had a loan which was to be paid next Friday, but we extended and allowed him to pay later.
>
> (FGD, Kisumu)

Associations can provide loans to their members through regular savings. It is from this pool of savings that members can access loans at a small interest rate, often at 10 percent compared to Kenya bank lending rate which stood at averagely 13.1 percent in 2018 (TradingEconomics, 2020). However, unlike banks or any other formal financial institutions which issue loans, the interest from loans offered by some associations is shared between members at the end of a certain period, as observed: "...we give loans but interest will be divided later" (FGD, Kisumu).

Yet, like formal institutions, late payment of loans attracts penalty or fines. This requirement does not discourage members from taking loans because income from fines is also shared amongst members. Some associations guard against defaulting by requiring a member requesting a loan to have a certain number of other members guarantee the loan.

Transformative measures

Using the Devereux and Sabates-Wheeler framework, one can argue that grassroot associations are structured to offer preventive and promotive social protection, while umbrella associations tend to offer transformative social protection which relate to representation and voice. In Table 7.3, a small proportion (1.5 percent) mentioned voice and representation as an important benefit of belonging to an association. This is because association members were largely sampled from grassroot type of associations.

The importance of voice and representation can be seen in the context of conflicts between city authorities and traders. As mentioned earlier, many of the umbrella associations have emerged to express grievances of traders as highlighted below:

> NISCOF was formed to reduce harassment and eviction of micro traders by city authorities. The Association negotiated for establishment of Muthurwa market for micro traders.
>
> (FGD, Nairobi)

Although strikes are a common way to express grievances in formal employment, only 17 percent of micro-traders had participated in a strike/protest. Since our research establishes that problems relating to the work environment are substantial, this finding suggests that protests may be an ineffective approach or that traders are employing other approaches to voice grievance. Other methods used to engage with the state include negotiation or soliciting help from political actors who perceive them as vote banks because of their large numbers. The main reason for engagement with the state is to improve their working conditions, as was noted in the following quote:

> ...we engage the County mostly on conditions of the market. Market must have shades and toilets. During rainy and sunny weather, traders need to

be protected from heat and rains. They also need to be provided with clean water because we cannot afford to buy water all the time. Regarding mobile traders, our Association has been negotiating for stoppage of harassment.

(KII, Nairobi)

Application of the power resource approach

This sub-section adopts the Power Resource Approach (PRA) as outlined in more detail in Chapter 1 of this book as a lens to identify power resources available to associations and how they are applied to achieve desired outcomes (Schmalz, Ludwig & Webster, 2018). Using the PRA lens, one can argue that micro-traders lack structural power which relates to the position of workers in the economic system. Structural power rests on the power to cause disruption (disruptive power) to such an extent that it interrupts productive use of capital. As mentioned, most association members (83 percent) have never participated in a strike or street blockage. This suggests that trader associations are not adequately organized to capitalize on such power; otherwise, they would make use of their large numbers.

Umbrella associations command greater associational power than grassroot associations. This is influenced by large membership, better organization, and leadership structure which enable them to play the role of voice and representation more effectively. In contrast, grassroot associations are often small, and their basis for organizing revolves around welfare and business. They are often inward-looking with conditions which keep others from joining. This trait makes it difficult to network with other associations.

While umbrella associations are more inclined to command associational power, they are weakened by inactive participation of members which reduces their material resources necessary for collective engagement with other actors. However, resource barriers can be reduced by forming alliances with other actors as similar to KENASVIT which has established networks with IDS Nairobi and donors to access material and technical support. MSEL, although it has a strong financial base, is inward-looking because it is organized around the creation of assets for members with hardly any alliances with other trader associations.

With respect to **institutional power**, umbrella associations are more inclined to appropriate this power when they have strong associational and societal power. Institutional power relates to the ability to use institutions such as legislation to advance own interests. Kenyan legislation has recently created structures that provide a platform for IWs to participate in making decisions that affect micro-traders mentioned by the chair of KENASVIT, who is a member of the board of MSEA. He uses his position to advance the interests of micro-traders and informal workers in general.

Societal power refers to power arising from cooperation with other social groups and organizations. There are two sources of societal power: coalition power and discursive power. Coalition power refers to having networks with other social actors at one's disposal and being able to activate these powers for mobilization and campaigns, while discursive power is the ability to interpret or frame burning issues to achieve societal power. Umbrella associations possess more of coalition power (as opposed to discursive power), especially those with interaction with other actors including politicians. An example is KENAVIT which has made alliances with local actors, transnational trader organizations, donors, and the state. Through these networks, it has been able to influence policy and acquire the resources necessary for lobbying. Lindell (2010) points out that the alliance between KENASVIT and Kenya Land Alliance was formed to voice concern on access to land from which IWs derive their livelihoods.

Although relations with politicians may grant societal power to associations to advance own interests, this power resource is time-limited because it can only be tapped when politicians are in position of power (KII, Nairobi). Political alliances potentially weaken group solidarity because of divergent voices within the group as is pointed out in following quote:

> ...every of our groups has a political friend. The unity that we once had has weakened. Politics has affected our membership. It has become a big problem.
>
> (KII, Nairobi)

The above discussion shows that PRA is useful to understand how power is formed or utilized by associations in voice and representation. However, there is difficulty in extending the PRA framework to grassroot associations whose focus is more inward-looking on welfare and business issues. The following section discusses some of the challenges of trader associations.

Associations' challenges

Trader associations face challenges of member participation which may limit their ability to deliver services to their members. Most associations have 'inactive' members who do not submit regular contributions. Inactiveness is most pronounced among umbrella associations who have a broader mandate of advocacy and lobbying and who need resources to mobilize. Thus, when more members fail to contribute, it affects activities of the association unless there is external support from donors, as is illustrated in the following quote:

> ...the office was closed last year because of members were not remitting the subscription fees. We were unable to pay the rent.
>
> (KII, Nairobi)

While there is power in collective engagement, sustaining solidarity between members or among groups is a challenge to some associations. One association noted that some members default on payment of loans. They choose to exit their association after receiving a loan (FGD, Kisumu). One informant confirmed that an association which he belonged to collapsed when two members took a loan and disappeared (KII, Kisumu). While loan protection insurance exists – a type of insurance designed to cover loan repayment and protect the insured from default – one participant pointed out that available products in the market are not suited for workers in the informal economy (KII, Kisumu). The above discussions show that weaknesses within groups may affect the abilities of the groups to deliver services (including providing link to formal insurance) to their members and to some extent lead to the 'demise' of groups.

Discussion

Access to associations

Entry conditions

Most trader associations are guided by constitutions that allow them to achieve their objectives. Although trader associations tend to offer a wide scope of social protection services (compared to formal insurance schemes) that could improve the livelihoods and welfare of their members, Table 7.3 shows that most trader associations (65 percent) have rules or conditions that prevent others from becoming members of these associations. Membership or entry to these associations depended on the nature of work (56 percent), geographic location (19 percent), gender (12 percent), social or family relations (5.8 percent), special conditions such as disability (2 percent), and others (5.9 percent).[10]

The findings were confirmed through observation and through discussion with association members and their leaders. Some associations comprised of women only while others were organized based on special needs, that is, persons with disabilities, persons suffering from HIV/AIDS, etc. Yet, others were based on the nature of work. Other than conditions, one can argue that membership fees and regular contributions may hinder entry to associations especially if these are high and incomes uncertain. This lock-out because of criteria reduces chances of accessing social protection from associations and further exposes non-members to poverty and need for social assistance.

Comparison between members and non-members

This section looks at the extent to which association membership is associated with social protection. It uses the list of relevant variables shown in Table 7.1 to compare association members with non-members with the object of identifying statistically significant differences between them. Table 7.4 shows that association members are significantly more likely to enrol or be covered by formal insurance schemes than non-members. As for gender, there is a higher share of men among non-members than among members suggesting that women are more likely to join associations. Association members tend to be older compared to non-members and are more likely to be married. In terms of

Table 7.4 Differences in key workers characteristics by association member status

	Not Member	*Member*	*Difference (N – U)*	*t-Value*
Formal insurance enrolment (health/pension)	0.15	0.35	–0.20	–3.40***
Health Insurance coverage	0.23	0.49	–0.26	–4.05***
Nairobi	0.63	0.52	0.11	1.62
Kisumu	0.37	0.48	–0.11	–1.62
Gender (male = 1)	0.60	0.35	0.24	3.68***
Married	0.55	0.68	–0.13	–2.00**
Local	0.28	0.29	–0.01	–0.18
Age	33.16	38.88	–5.71	–3.94***
Mean daily earnings (current USD)	7.94	13.16	–5.21	–3.35***
Assets (house and/or land)	0.18	0.26	–0.08	–1.39
Primary incomplete	0.18	0.15	0.04	0.73
Primary complete	0.33	0.45	–0.12	–1.71*
Secondary or above	0.48	0.40	0.08	1.15
Training course	0.02	0.06	–0.04	–1.26
Training on job	0.22	0.19	0.03	0.49
Self-taught	0.76	0.75	0.01	0.15
Wageworker	0.14	0.00	0.14	4.64***
Own-account	0.84	0.79	0.05	0.96
Micro-business	0.02	0.21	–0.19	–4.14***
Observations	223			

Source: Author's elaboration based on project survey data.
Notes: *** <0.01, ** <0.05, * <0.1.

incomes, members of associations are more likely to have higher incomes compared to non-members. This matches the finding that members are more likely to be micro-enterprise owners who have higher earnings, whereas non-members are dominantly wage-workers with lower earnings. In sum, the findings suggest that members of associations are more likely to be older married women who are micro-enterprise owners with higher earnings and more likely to enrol themselves and their families in formal insurance schemes.

The findings thus suggest that associations play a crucial role in the adoption of formal insurance among their members. Discussion with association members and their leaders confirms this finding. Some associations had indeed played a crucial role in registration of members to NHIF as evidenced in the following quote:

> ...All members of my group are registered for NHIF and are contributing KES 200 every month through the group. Every week, the group takes a sum of KES 7000 to the NHIF. if a member is unable to pay the KES 200 the group takes the money from the members' savings and pays for him or her. The member who is being paid for is considered to have a debt of the group which he or she must pay.
>
> (FGD, Kisumu)

The above quote not only shows how some associations encourage enrolment but also shows how a group can prevent default among their members. Apart from this example, there are no other cases of associations paying contributions for their members. Nairobi Informal Sector Confederation (NISCOF) which is an umbrella association made efforts to register their members in the NHIF scheme but experienced a challenge paying premium as a group and opted for each member to make their own payment.

Comparison between the enrolled and the non-enrolled members of formal insurance schemes

This section examines the differences between the enrolled and the non-enrolled members of formal SI schemes. In line with Table 7.4, Table 7.5 shows that there is a higher share of association members enrolled in formal insurance schemes compared to non-members. The formal insurance members are more likely to be married, of a higher age, and are more likely to belong to worker-related associations offering loans. They are also more likely to be better off

Table 7.5 Differences in key workers characteristics by social insurance enrolment

	No Social Protection	*Social Protection*	*Difference (N – U)*	*t-Value*
Association member	0.54	0.79	−0.24	-3.40***
Sacco/vicoba/chama	0.40	0.43	−0.02	-0.34
Work related association	0.09	0.26	−0.18	-3.51***
Women/youth/religious	0.06	0.10	−0.04	-1.14
Work-related	0.09	0.08	0.01	0.25
Loans	0.44	0.67	−0.23	-3.08***
Voice and representation	0.01	0.02	−0.01	-0.72
Social cushioning	0.00	0.02	−0.02	-1.64
Nairobi	0.57	0.54	0.03	0.44
Kisumu	0.43	0.46	−0.03	-0.44
Male	0.42	0.52	−0.10	-1.40
Married	0.57	0.80	−0.24	-3.31***
Local	0.26	0.34	−0.09	-1.26
Age	35.54	39.59	−4.05	-2.50**
Mean daily earnings (current USD)	9.74	14.81	−5.07	-2.96***
Assets (house and/or land)	0.20	0.31	−0.11	-1.70*
Primary incomplete	0.19	0.10	0.09	1.57
Primary complete	0.42	0.36	0.06	0.80
Secondary or above	0.40	0.54	−0.15	-1.97**
Training course	0.04	0.07	−0.03	-0.92
Training on job	0.17	0.28	−0.11	-1.76*
Self-taught	0.79	0.66	0.13	2.09**
Wageworker	0.07	0.02	0.05	1.52
Own-account	0.84	0.72	0.12	2.00**
Micro-business	0.09	0.26	−0.17	-3.33***
Observations	223			

Source: Author's elaboration based on project survey data.
Notes: *** =0.01, ** =0.05, * =0.1.

in terms of education, with higher incomes and assets compared to those not enrolled in formal insurance schemes. When it comes to employment status, formal insurance members are more likely to be micro-enterprise owners, while own-account are less likely to enrol in formal insurance schemes This is to be expected, as was noted earlier that micro-enterprise owners had higher mean daily earnings of USD 20 compared to own-account workers with a mean daily earning of USD 10.

Conclusion

This chapter has examined the socio-economic characteristics of traders, the characteristics of associations including their challenges, and the extent to which they influence access to formal and informal social protection services. It has shown that many micro-traders still do not have access to health insurance and pensions offered by the state. It has also revealed that health cover is more important to traders than pension cover and that reasons for not signing up (or defaulting) are influenced by factors related to socio-economic characteristics such as low and irregular incomes and factors related to service provision such as cost of premium and cumbersome procedures. This suggests the need to redesign the NHIF scheme to make it attractive in terms of cost and to adjust the premium payment plan to the realities of informal work characterized by low and irregular incomes and long working hours.

The low uptake of formal insurance among IWs also begs the question of how non-enrolled members of formal insurance schemes cope without insurance and whether there are any other existing models for provision of social insurance to IWs. This chapter shows that trader associations have developed their own forms of social protection to support themselves, albeit not adequate in terms of meeting all needs. Using the Devereux and Sabates-Wheeler framework, this chapter has shown that social protection models of associations are designed to avert risks and shocks touching on associational life and livelihoods (preventive social protection), are useful in resource mobilization, as well as, in averting risks and shocks that threaten their livelihoods through voice and representation (transformative social protection) and through offering timely loans and training to members (promotive). However, these associations have weaknesses which affects the delivery of social protection. One main weakness of grassroot level associations is that most of them are small and are guided by regulations that restrict membership thus locking others out. Due to their small size, their resources are limited to meeting immediate needs.

This chapter has also shown that a higher share of association members enrols in formal insurance schemes than non-members. Indeed, there are some associations which enabled their members to access formal health insurance through provision of information and necessary support including making of regular payments.

Although the PRA framework reveals that there is strength in coming together to overcome challenges in the sector, this chapter has shown that most of the existing grassroot associations are 'inward-looking' which limits them from forming alliances with other associations. Furthermore, umbrella associations are mainly driven to action in cases of conflicts between micro-traders and city authorities. A few cases of active engagement are made possible through support of external institutions and actors, including politicians who provide resources needed for collective organizing and technical support.

Notes

1 Micro-trading here means activities of a person who owns or operates micro-enterprises with maximum employees and persons employed in such enterprises.
2 The average USD to KES rate used throughout the chapter is KES 101.2, rate from June 2018 to December 2018.
3 'Jua kali' is a Swahili word meaning 'hot sun'. The term reveals that informal activities are conducted in hot sun.
4 In addition, the UAC Act establishes a Board to manage the affairs of cities and town.
5 'Chama' is Kiswahili for groups formed by individuals with common objectives providing services like loans and savings, including welfare services.
6 This information is based on an interview with current KENASVIT Chairperson and with leader of Kisumu Informal Trader Economic Support (KITES).
7 This information is derived from interview with a representative of the Association.
8 SACCO activities are regulated under SACCO Societies Act, 2018.
9 Information was obtained from the Chairperson of the group.
10 On the second response, geographic location was the main condition (53.1 percent), followed by gender (37.1 percent), and others.

References

Brown, A. (2006) Challenging street livelihoods. In: Brown, A. (ed.) *Contested space: Street trading, public space and livelihoods in developing cities*. Warwickshire, ITDG Publishing, pp. 3–15.

Chen, M., Roever, S. & Skinner, C. (2016) Editorial: Urban livelihoods: Reframing theory and policy. *Environment and Urbanization*. 28 (2), 331–342. Available from: doi:10.1177/0956247816662405.

Devereux, S. & Sabates-Wheeler, R. (2004) *Transformative social protection*. IDS Working Paper 232. Brighton, Institute of Development Studies.

DN (17 October 2017) Sonko abolishes charges for city small-scale traders. *Daily Nation*. Available from: https://www.nation.co.ke/kenya/news/sonko-abolishes-charges-for-city-small-scale-traders--458052.

Hart, K. (1973) Informal income opportunities and urban employment in Ghana. *The Journal of Modern African Studies*. 11 (1), 61–89.

ILO (1972) *Employment, incomes and equity: A strategy of increasing productive employment in Kenya*. Geneva, International Labour Office.

ILO (2002) Report VI, Decent work and the informal economy, Sixth item on the agenda. In: International Labour Organization. *International Labour Conference 90th Session*. Geneva, International Labour Office.

ILO (2017) *World employment and social outlook: Trends 2017*. International Labour Organization. Geneva, International Labour Office.

KNBS (2016) *Micro, small and medium (MSME) establishment survey: Basic report 2016*. Nairobi, Kenya National Bureau of Statistics.

KNBS (2018) *Kenya integrated household budget survey 2015/2016*. Nairobi, Kenya National Bureau of Statistics.

Lindell, I. (ed.) (2010) *Africa's informal workers: Collective agency, alliances and transnational organizing in urban Africa*. London, Zed Publications.

Maloney, W.F. (2004) Informality revisited. *World Development*. 33 (7), 1159–1178.

Mitullah, W.V. (2003) *Street vending in African cities: A synthesis of empirical findings from Kenya, Cote D'Ivoire, Ghana, Zimbabwe, Uganda and South Africa*. Washington, D.C., World Bank.

NCLR (2019) *Urban and cities (amended) act*. Nairobi, National Council for Law Reporting.

Racaud, S., Kago, J. & Owuor, S. (2018) Introduction: Contested street: Informal street vending and its contradictions. *Journal of Urban Research*. 17–18. doi.org/10.4000/articulo.3719

ROK (1992) *Sessional paper no. 2 of 1992 on small enterprise and jua kali development in Kenya*. Republic of Kenya. Nairobi, Government Printer.

ROK (2005) *Draft sessional paper on development of micro and small enterprises for wealth and employment creation for poverty reduction*. Republic of Kenya. Nairobi, Government Printer.

ROK (2019) *The street vendors (protection of livelihood) senate bill, 2019*. Republic of Kenya. Nairobi, Government Printer.

Schmalz, S., Ludwig, C. & Webster, E. (2018) The power resources approach: Developments and challenges. *Global Labour Journal*. 9 (2), 113–134.

Skinner, C. (2008) *Street trade in Africa: A review*. Working Paper No 5. WIEGO.

TradingEconomics. (2020) *Kenya bank lending rates*. Available from: https://tradingeconomics.com/kenya/bank-lending-rate.

WIEGO (2019) *Extending social protection to informal Workers*. WIEGO. Available from: https://www.wiego.org/publications/extending-social-protection-informal-workers.

8 Social protection and informal construction worker organizations in Tanzania

How informal worker organizations strive to provide social insurance to their members

Aloyce Gervas

Introduction

Tanzania is on a path of moving from being a low-income to a middle-income country with industrialization as a key priority. During the last ten years, Tanzania has experienced relatively high economic growth, averaging 6–7 percent a year (World Bank, 2018). Among the contributors to growth, the construction sector has played a key role contributing 13.2 percent to GDP as of the first quarter of 2019, making it the largest driver of GDP in the country (NBS, 2019). According to Mkenda and Aikaeli (2015), the growth of construction activities in Tanzania is a result of two factors, the increase of government budget allocation to infrastructure development and the increase of private investors in the real estate business and cement production.

Most construction activities in Tanzania, as it is in other developing countries, remain labour-intensive requiring less technical skills, thus allowing unskilled and low-skilled labour to enter the sector (Wells & Jason, 2010). A substantial part of construction activities within the country is undertaken by informal micro and small private entrepreneurs (Wells & Jason, 2010; NBS, 2013), who are subcontracted to perform construction work by large contractors who bid and win the tenders. As a result, most workers in the construction sector in Tanzania are working informally on different projects from high-rise buildings to bungalows, roads, and bridges, among others. Due to the nature of these activities, workers are subjected to various occupational risks and have very minimal or no social protection. As noted by Jason (2007), 95 percent of serious accidents in construction sites involve Informal Construction Workers (ICWs) who mostly perform the work without contracts or with temporary/verbal contracts, leaving the liability for ensuring health and safety on their shoulders.

The construction sector in Tanzania contributes to about 25–45 percent of all the occupational fatalities in the country (Mrema, Ngowi & Mamuya, 2015), and it is the leading sector in terms of fatalities with a rate of 23.7 percent. To ensure that workers are safe in construction sites, Tanzania has established various

DOI: 10.4324/9781003173694-8

institutional machineries that regulate the construction sector, register the contractors, and provide different rules and regulations that govern construction activities in the country. These institutions include the Contractors Registration Board (CRB), the Occupational Safety and Health Agency (OSHA), and the Workers Compensation Fund (WCF) among others. These institutions are mandated to oversee the operations of different activities within the sector ensuring that policies, rules, and regulations are adequately followed; yet, many ICWs continue to operate without well-defined rights and in precarious environments without social protection measures (ILO, 2017).

The construction sector in Tanzania falls under the Tanzania Mines, Energy, Construction and Allied workers union (TAMICO) which deals with workers' rights in the sector. The trade union currently serves only formal employees in the construction sector but is open to own-account or informal workers who contribute the union fees from time to time. However, TAMICO has shown a significant deficiency in dealing with workers in the informal sector (Jason, 2007; Wells & Jason, 2010) since there are no employers for the informal sector workers and therefore making it difficult for collective bargaining. An established construction sector union for informal workers (TAICO) was formed as part of a UNDP project implemented between 2002 and 2008, where a total of about 43 groups of informal workers joined and were expected to operate under the umbrella of TAMICO. However, the association collapsed after the UNDP funding was over, and TAMICO has not made efforts to revive the association (Jason, 2007).

This chapter analyzes how ICWs organize their associations, and how associations facilitate the provision of social protection measures to their members. There is limited knowledge on ICWs associations, and this chapter aims at filling the gap as well as contributing to the body of knowledge on social protection.

Methodologically, the chapter draws on both quantitative and qualitative methods, and the respondents included in the study were both wage workers and own-account workers, employed directly by construction site managers or indirectly via intermediaries such as gang leaders working on large and medium building sites.[1] The study focused on public and private buildings excluding residential housing sites. A total of 212 respondents across two regions (Dar es Salaam and Dodoma) were interviewed. However, after data cleaning, a sample of 205 respondents was used for analysis. As for the qualitative data, 14 key informants who are leaders of worker associations were interviewed, seven from Dar es Salaam, five from Dodoma, and two trade union leaders. In addition, four focus group discussions with members of worker associations involving six to eight people were conducted across the two regions.

Apart from this introductory part, Section two is key worker characteristics which provides a brief synopsis of key characteristics of respondents involved in the study highlighting their various characteristics such as education, age, and construction skills. Section three is association landscape for informal construction workers introduces the associational landscape followed by brief histories of associations in the study areas discussing reasons for formation, membership criteria, and services offered. Section four is powers of informal construction workers associations that discusses various powers of associations such as associational, institutional, societal, and structural powers. Section five is services offered by associations and it covers the various services offered by associations followed by a discussion in Section six titled access to associations where entry conditions, and barriers to entry are discusses. Section seven is access to formal social insurance and focuses on access to formal social insurance of informal workers, detailing those enrolled and the reasons for not enrolling followed by a conclusion.

Key worker characteristics

Table 8.1 presents some of the key characteristics of the 205 ICWs interviewed across the two geographical locations, 44 percent from Dar es Salaam and 56 percent from Dodoma. Almost all workers (99 percent) were male with an average age of 36 years and 68 percent were married. In terms of education, 56 percent of the workers had completed primary education, while 23 percent had completed secondary education. In terms of training, 39 percent of the workers got the construction skills through being self-taught, while 35 percent acquired the skills on the job and 26 percent attended a training course in construction skills.

Table 8.1 Key worker characteristics

	All		*Dar*		*Dodoma*	
	Mean	*SD*	*Mean*	*SD*	*Mean*	*SD*
Member*	0.19	0.40	0.11	0.32	0.26	0.44
Social insurance enrolment	0.18	0.39	0.14	0.35	0.21	0.41
Health insurance coverage	0.17	0.38	0.12	0.33	0.21	0.41
Gender (male = 1)	0.99	0.10	0.99	0.11	0.99	0.09
Age	36.38	9.43	36.21	9.06	36.51	9.74
Married	0.68	0.47	0.63	0.48	0.72	0.45
Local born	0.40	0.49	0.26	0.44	0.52	0.50
Mean daily earnings (current USD)**	14.76	22.30	10.72	13.64	17.92	26.87
Assets (house and/or land)	0.50	0.50	0.41	0.49	0.57	0.50
Primary incomplete	0.07	0.26	0.07	0.25	0.08	0.27
Primary complete	0.56	0.50	0.53	0.50	0.58	0.50
Secondary and above	0.37	0.48	0.40	0.49	0.34	0.48
Professional training course	0.26	0.44	0.23	0.43	0.28	0.45
Training on job	0.35	0.48	0.26	0.44	0.43	0.50
Self-taught	0.39	0.49	0.51	0.50	0.30	0.46
Wage-worker	0.32	0.47	0.40	0.49	0.26	0.44
Own-account	0.50	0.50	0.51	0.50	0.50	0.50
Micro-business	0.18	0.38	0.09	0.29	0.24	0.43
Observations	205		90		115	

Source: Author's elaboration based on the project survey data.

Notes: * Based on the random sample consisting of 151 workers. ** The median wage is USD 9.0 overall, USD 7.7 for Dar es Salaam, and USD 10.8 for Dodoma.

Informal workers in the construction industry generally conceptualize themselves as both workers and businessmen, since once they are hired in big construction projects, they work under an employer, but sometimes, they secure their own assignments and thus work as own-account. Often, employers do not really care about their identity but rather concentrate on their specific skills and how fast they can fulfil the assigned tasks. From the survey data, the majority of ICWs in Tanzania regard themselves as own-account workers (50 percent), while 32 percent regard themselves as paid workers, and around 18 percent indicated that they were engaged in micro-business with some variations across the two locations.

The survey also revealed that, of all the respondents interviewed, only 19 percent had enrolled in various forms of worker associations though this differed across locations with association membership being 26 percent in Dodoma and 11 percent in Dar es Salaam. The percentile in Dodoma is higher because the majority are local born, and it is easier for them to organize unlike in Dar es Salaam where the majority have moved in to search for jobs. As for social insurance (SI), overall 18 percent of workers contributed to formal schemes, out of which 11.2 percent were contributing to a health insurance, 2.9 percent are contributing to a pension fund, and 3.9 percent to other kinds of insurances. Interestingly, the overall enrolment was higher in Dodoma at 21 percent compared with Dar es Salaam, where 14 percent of workers are enrolled. As discussed above, the difference between Dodoma and Dar es Salaam is related to place of domicile, as the majority of those who live in Dodoma are local born. This is most probably related to the existence of the Community Health Fund in Dodoma as discussed below. As for actual health insurance coverage, the figures are more or less in line with SI enrolment.

ICWs in Dar es Salaam and Dodoma have some distinctive differences. For instance, in terms of origin, 52 percent of the workers in Dodoma were local born, while in Dar es Salaam, it is lower at 26 percent. This could be due to rural–urban migration, where workers move from their home areas to the main city in search for jobs. Also, the mean income which is around 15 USD is lower in Dar es Salaam compared to Dodoma because Dar es Salaam has many construction workers competing for the same construction sites, and workers are willing to work on prices decided mostly by the site engineers and workers hardly negotiate the terms. Furthermore, the nature of the respondents interviewed in Dar es Salaam had more helpers who earn less than the masons who hire them while Dodoma had more masons who hire the helpers. It is also revealed that, in Dodoma, construction workers are few, and thus, they dominate the construction works available. Thus, the median wage is lower than the mean due to the right skew of the distribution with a few heavy earners at the top pulling up the mean. For this reason, the median is a more accurate measure of the general wage level. Finally, in terms of ownership of assets, 50 percent of the respondents had assets such as land or a house, slightly higher in Dodoma at 57 percent compared to 41 percent in Dar es Salaam.

Associational landscape for informal construction workers in Tanzania

Informal construction workers (ICWs) as compared to informal workers in other sectors such as trade and transport in Tanzania have lower representation in associations. Data from other sectors reveal that, in trade, 34 percent are members of associations, while, in transport, it is 50 percent compared to 19 percent in construction. The notion of coming together and forming associations is somewhat a new phenomenon for the ICWs. This may partly explain the relatively low share of association members as revealed in the survey. Table 8.2 below provides insight into the types of associations ICWs belong to.

Table 8.2 shows that, in terms of association type, 22 percent of ICWs were in either Saccos/Vicoba or Chama, 76 percent were in worker associations, and 2 percent in other associations such as religious, youth, and women, while, in benefit type, the study revealed that 54 percent were in specific work-related associations, for example, masonry, carpentry, or plumbers. In addition, 19 percent receive social cushioning, and only 20 percent of the members have enrolled in social insurance which is average across all workers. Association members also receive loans, 27 percent, and 60 percent pay membership fees to associations.

The ICWs in Tanzania perform their construction activities very independently. Most workers meet at construction sites waiting to be hired independently by

Table 8.2 Key worker characteristics, association members

	All		*Dar*		*Dodoma*	
	Mean	*SD*	*Mean*	*SD*	*Mean*	*SD*
Social insurance enrolment	0.20	0.41	0.19	0.40	0.21	0.41
Health insurance coverage	0.22	0.41	0.22	0.42	0.21	0.41
Association type						
Sacco/vicoba/chama	0.22	0.41	0.07	0.27	0.29	0.46
Worker association	0.76	0.43	0.89	0.32	0.70	0.46
Other association	0.02	0.15	0.04	0.19	0.02	0.13
Benefit type						
Work-related	0.54	0.50	0.74	0.45	0.45	0.50
Social cushioning	0.19	0.40	0.07	0.27	0.25	0.44
Voice and representation	0.00	0.00	0.00	0.00	0.00	0.00
Loans	0.27	0.44	0.19	0.40	0.30	0.46
Barriers	0.70	0.46	0.70	0.47	0.70	0.46
Fee	0.60	0.49	0.67	0.48	0.57	0.50
Observations	83		27		56	

Source: Author's elaboration based on the project survey data.

the site engineers. Associations in construction, therefore, have been lagging behind compared to other sectors.

Histories of informal construction workers associations[2]

To understand ICW associations, two association cases are presented to provide a holistic view of different types of organizations, their formation, and organizing principle – including membership structure and contributions, how they have been able to support their initiatives, success stories or lack of, as well as the challenges they face. The associations are similar in some respects, but different along other dimensions such as how they are organized, services offered, challenges, as well as their potential for growth. In general, they are quite representative for the associations encountered in this study.

Upendo Group[3]

Upendo group (translated as Love group) is an association of informal construction workers which was established in 2015 in Dar es Salaam. It is an informal group, not yet registered but operating in the Pugu area with 42 members. During the research, they were in a process to apply for a license to register the association formally. The group was established with the intention of seeking job opportunities together, seeking recognition from the local authorities, and attracting employers such as engineers and house owners for job opportunities. The group has a hierarchical leadership structure, with a Chairman, Vice Chairman, Secretary, and Treasurer who are elected by the members every five years.

Membership to this group is open to all ICWs residing in Dar es Salaam. Requirements to become a member of this association include filling in the application form and being interviewed by the Secretary and Chairman. According to the Chairman of the association, the group operates by the highest standards of Integrity, Discipline, Honesty, and Love.

Discussions revealed that the group had grown to perform various projects including building schools, hospitals, and dispensaries for the local government after getting the tender together with other informal construction workers. It was noted that it was the responsibility of the group members to seek for jobs and bring them to the group so that the group assigns workers to the different tasks. The Chairman revealed that other employers and site engineers approach the group and request for their labour.

The chairman further reported that it is the responsibility of the association to take care of the members when they fall into trouble. The association represents them to the authorities or when they are demanding their rights such as payments. Also, in case, a member is sick, or a close family member is, the group contributes to solve the challenge. A certain amount is drawn from the account depending on the balance while the remaining amount is contributed by the members. They do not have a limit. Cushioning mechanisms during difficult times is the most important achievement of the association and something that the members did not think about when they first organized. Each member contributes TZS 5,000 every month. Another source of income is generated from the work that is done; a certain percentage of the individual profit is deducted and retained by the association. For example, one association leader noted:

> Understanding that there is potential risk of being injured while working, we have built a habit of putting some money as a reserve to take care of such situations. For example, when our fellow member suffers an injury while at work, we would be compelled to take care of him. So, to address that noble mission we decided every member to contribute TZS 10,000/= every month so that, when our fellow member feel unwell, we take him to the hospital and take care of his needs.
>
> (KII respondent)

Although they have organized and made certain achievements, they feel that they are yet to be recognized properly, they cannot access external loans, perform larger jobs, nor have a strong voice because they are not formal.

Dodoma Construction Workers Association 'Chama cha Mafundi Ujenzi Dodoma'[4]

This association was established in 2008 and has been operating in the Dodoma region for more than 11 years. Founded by four members and now with a total number of 11 members who are all ICWs. The association is not registered, and its operation relies on the availability of jobs. The association was established to secure jobs and work as a team to accomplish construction assignments. The association has leaders who were appointed when it was formed, and to date, they have not done any election of new leaders.

Interested members pay a fee before joining, and they have a monthly contribution of TZS 10,000. The association is open to all construction workers in Dodoma. There are no restrictions to join the association if one is a construction worker and able to perform their duties. The Chairman noted that it was difficult in the past to secure a job. Most workers were from different areas, and sometimes, the working environment was not harmonious once a job is secured. He further observed that the establishment of the association had improved the synergy of members, and there have been success stories of members working together, including jointly deciding how to perform the work activities and assisting each other if need be.

According to the Chairman, the association represents the members towards the authorities when they face a challenge that requires support such as harassment from the employers or authorities. They do also provide support services to members including loans when a member faces a challenge like a family member is sick or passes on.

He noted that they have been facing challenges of members moving to other towns and regions. This has been a reason for low membership and hindrance to organizing, including inability to meet and conduct elections. The entrance fee and monthly contribution was also viewed by the Chairman to be a contributing factor to the exit of members. There are cases when construction workers do not get jobs at all and are unable to contribute. The association feels isolated due to informality, the workers are not recognized, and they feel that they do not have a place to air their issues to get attention. They also see opportunities to grow, but again, they do not have capital or capacity to borrow or buy work tools and equipment.

Overall, the association is not very active, and members are equally not active. It seemed that the leaders have not been very keen to attract more construction workers.

Powers of informal construction worker associations

The Power Resource Approach (PRA) as it has been used by researchers (Wright, 2000; Silver, 2003; Schmalz, Ludwig & Webster, 2018) contends that, with different power resources (associational, structural, institutional, and societal powers), workers are able to collectively mobilize and advance their issues or interests. This section discusses how this framework applies in the context of ICWs.

Associational power

Associational Power entails workers in the informal sector coming together and forming collective organizations that address their common interests. The ICWs associations in Tanzania have demonstrated the presence of associational power in terms of coming together for common interests, infrastructural resources, member participation, and organizational efficiency. Although only 19 percent of ICWs are members of associations, during interviews, it was noted that the association have assisted the informal workers in various ways.

Associational power for construction associations differs a bit from the traditional understanding of associational power as discussed in the literature (e.g. Schmalz, Ludwig & Webster, 2018) where it is expected that the collectivizing of the workers would give them more bargaining power against the employers, for example, pressuring employers into providing social protection measures or salary increase. However, ICWs use association power more to tackle issues that directly affects them such as welfare issues and recognition by the authorities as well as for joint action to improve their job opportunities and increase efficiency and productivity.

An association such as Upendo Group has been recognized by the local government authorities to the extent of being awarded small government tenders such as building or renovation of hospitals, schools, and dispensaries. This is an example of organizational efficiency as shared by one of the leaders:

> We were not recognized, and we got so many challenges in getting jobs, thus we organized with my fellows and we started a group so that it can be easy to be recognized and get jobs.

The same was observed by Sinza Kijiweni Construction Group that came together and paid for an office space through a fee to the local government. Various potential employers with construction projects come to their office location and hire construction workers.

The associations have also been able to facilitate various social welfare issues affecting their members such as life tragedies as sickness, death, financial support/loans for schools' fees. These are forms of informal social protection schemes that, through associational power, members have been able to access. They happily operate in their offices and are able to receive visitors as reflected in Photo 8.1 below.

Those who are operating solo in the sector must face life contingencies such as sickness, death, or financial difficulties by relying on other mechanisms. In a discussion with one leader of the Nkurabi Construction Association in Dodoma, the leader noted that a key strength for them was the responsibility of the association to stand for their members during such moments.

Representation and voice did not come out among the benefits in the survey for those who are members of associations. However, from some of the Focus Group Discussions and Key Informants, it was noted to be an important benefit. Some of the associations such as Keko Carpentry, Sinza Kijiweni construction association in Dar es Salaam through associational power, have managed to represent all the carpenters and masons among others to raise voices in demand for their rights including workspaces or against harassment by authorities. This is

Photo 8.1 Members of Sinza Kijiweni Construction Group posing for the picture at their office location in Dar es Salaam.

also evident in other groups, such as Msigani construction group and Mpunguzi construction group in Dodoma where they had both managed to represent members although in a very low scale such as demanding to be paid their money after they had worked for the local government.

Member participation has also been a key importance for the associations. It is evident where members have respected the association rules and regulations and participated in meetings organized by the leaders, such associations have proved to be progressive unlike those where members are not fully participating in the association issues. This is backed up by association leadership efficiency in organizing members. Examples could be drawn from Dodoma Construction workers, although present for over ten years, they still have few members, and low participation in association meetings. Upendo group on the contrary is only five years, but has attracted over 42 active members who are fully participating in association activities.

It was also interesting to note that, most of the older associations were not as strong as the younger associations in terms of organizing, but rather, they seemed to be dormant. This scenario could perhaps be associated with younger associations setting up with a specific purpose, whereas old ones, once the purpose was met such as being known by the potential clients, the organizational power fizzled.

Institutional power

Institutional power is slowly coming out as an area among informal construction workers as some associations try to formalize the relations between the informal worker associations and the authorities through performing tendered government projects. This was specific to one association (Upendo Group) where we see an association performing construction projects linked directly with the local government authority. Local government authorities seem to appreciate the services offered by this construction association as noted by a KII:

> After coming together at least we were recognized even local government offices recognized us. If the local government have any construction job such as building school, they call us, and we get paid by the government.
>
> (KII, Dar es Salaam)

However, most of these associations are mainly for joint job seeking and welfare for the workers. The primary goal had not been to institutionalize their activities because that seems to be a dream for the future.

Structural power

For ICWs, structural power is not evident. Workers, due to the nature of their activities, are not able to bargain due to tightness of the labour market and qualification demands of the employers prior to employment. This is also the case when workers' rights are infringed, there is little or no possibility for them to demand their rights. As a key informant in Dodoma explains:

> Sometimes we enter into agreement with a company boss to perform a specific task, but at the end the employer decides not to pay you as agreed claiming that your work was substandard, sometimes you might not be paid at all. You decide to quit but others will continue working, we need the money for survival.
>
> (KII, Dodoma)

The KII point was corroborated with the survey data. For example, when members were asked if they had ever participated in blocking construction sites, strike or work stoppage, 91.6 percent said no and only 8.4 percent said yes. This suggests that the associations are not yet strong in teaming up and demanding for their rights, although they may be using other negotiation tactics.

As noted from Keko Carpentry association and Dodoma Construction Workers Association, members receive support from the groups to take care of their injuries after they have failed to secure their rights from the employers, instead of going to the employer and demanding for employment injury support.

Societal power

Regarding societal power, it is again not evident in Tanzania unlike in Kenya, for example, where construction associations have managed to work with NGOs and other international platforms. Most construction associations are inward looking, concentrating more on their immediate needs. Since the employment relation is not very transparent and the associations are not under any umbrella organization or trade union, they tend to take care of their issues within the associations.

The findings revealed that associational power seemed to be the dominant form of power amongst ICWs both in Dar es Salaam and Dodoma. For these workers, organizing meant coming together to search for jobs and cushioning themselves during depriving life situations.

Services offered by associations

The following section draws on the social protection framework by Devereux and Sabates-Wheeler (2004), as outlined in Chapter 1, focusing on preventive, promotive, and transformative measures.

Preventive measures

Preventive measures are a type of social protection as derived from the Sabates-Wheeler framework that includes formal social insurance for economically vulnerable groups for coverages such as health insurance and pension. This type of social protection seems to be unavailable for informal construction workers due to the nature of their jobs. The construction sector is very hazardous, and workers are prone to accidents. On occupational health and safety, the findings from the survey revealed that at least 66.8 percent of the respondents had been injured or suffered an occupational illness.

Figure 8.1 shows that, of the injured workers, 69 percent had to pay for their medical bills, and only 14 percent of the bills were paid by an employer. There is a small percentage (4.4 percent) who used their insurances while others depended on relatives, work colleagues, and associations. Chances that the employer would take care of the injured person when the accident occurs to full recovery were very limited. This came out as a key concern of workers during the interviews. They noted that the employer would only give you first aid treatment or take part of the initial hospital bill and afterwards leave you to friends and relatives.

Safety for workers is guided by the Tanzania Occupational Safety and Health Act (OSHA) Number 5 of 2003 and other legal mandates such as Building and Construction Rules, 2015 among others. The findings from this study however reveal that workers are still complaining about issues regarding health and safety at work. The perception from workers is that most contractors (employers) are not concerned about their health and safety.

In terms of confinement to the law and regulations, OSHA's definition of an employee, it is whoever works in the building whether on short- or long-term contract. OSHA, however, does not recognize people without contracts. Of those who perceived themselves to be paid workers, 39 percent had no contract, while 24 percent had informally agreed contracts, and 10 percent had contracts of more than six months. Those who perceived themselves to be own-account workers most likely offer their labour at their own risk in the construction sites (see Photo 8.2). The findings are also in line with a study done by Mkenda and Aikaeli (2015) in Tanzania where the results showed that about 82 percent of

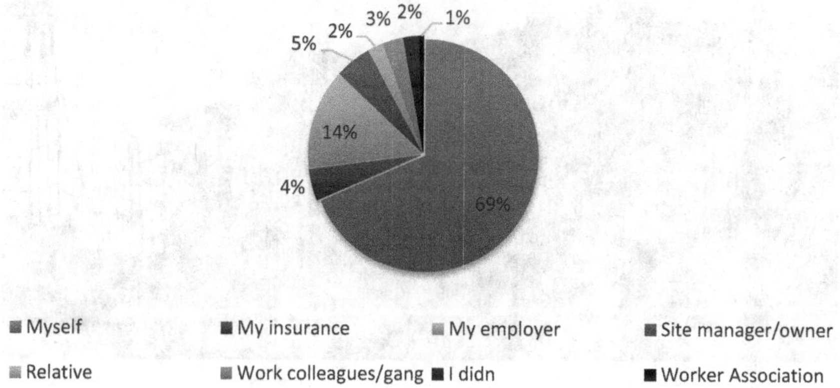

Figure 8.1 Responsibility of medical bills after injury.

Source: Author's elaboration based on project survey data.

formal firms use informal construction workers for all their projects and 76 percent of all those surveyed did not sign any contracts.

All the associations covered in the research were very strong in the provision of preventive measures as an aspect of informal social insurance. It came out strongly that the associations have succeeded to cushion the members during difficult times, as observed by one member:

> We have one hope only, and that is our associations. That is why we call our group, 'Our Hope'. When we get jobs to perform, we inform our leader and then contribute to the association a certain percentage depending on the amount received but not less than TZS 5,000. This contribution is the one that would help me or others during difficult moments. That is how we assist each other.
>
> (FGD, Dar es Salaam)

It was revealed during discussions that some members may not contribute willingly to the association, but they could be the first ones to seek for help when in trouble. Therefore, all the associations have a mechanism in place to track the contributions of each member and how that member has socially engaged with others when they get problems. For example, Sinza construction group has a unique mechanism for tracking members' contributions. Once a member gets a problem such as an accident resulting from work, sickness, or death of a family member, a special book (Photo 8.3) is bought for that specific member. All members contribute, and their names and amount contributed are recorded in the book which acts as a receipt.

Photo 8.2 Dar es Salaam. A picture showing a group of informal construction workers working at height without any form of PPE while the site housekeeping is not well arranged.

Other groups such as Ten brothers, Love group, Msigani group, and others have one counter book, where members' contributions are recorded. The amounts contributed vary from one group to another, for example, group members from the Nguvu Kazi and the Accident and Death Group noted the following:

> If you have lost your beloved ones (wife, kids, father and mother) i.e. only these four categories of people, as a group we provide TZS 500,000 and every member contributes TZS 30,000 on top of that.
>
> (FGD, Dar es Salaam)

Other associations have different mechanisms of support. It was noted that some associations could support any family member who has been living with a member for years in their household. This is different from formal SI measures which only

Photo 8.3 Special textbook for contributing to members who got challenges from Sinza Construction Group, Dar es Salaam.

apply to a certain number of named dependents. With this informal SI measure, the informal workers are not left to take care of life burdens individually.

It was also interesting to note that, once a member had an occupational accident and he cannot work for some time, some associations, in addition to covering the medical bills, would go on assisting the member by sending some monetary allowances every week until the member is able to resume work. This support is not a loan to be returned, but like a sick leave allowance as noted by one FGD participant (Photo 8.4):

> I got an accident when I was working at height in a construction site and I broke my leg after falling from a scaffold, my employer never supported me and thus my association took the liberty of taking care of me. After I was released from hospital, my association continued to support me. Every Sunday they bring TZS 30,000 for me and my family, it is now six months.
>
> (FGD Respondent)

However, it was noted that the ICWs associations had not been active in terms of facilitating members to join the formal insurance schemes. Among others, members noted the reasons for being in the infant stages of formation of the

Photo 8.4 An informal construction worker who broke his hip after a work accident using a self-bought crutch to walk six months after he got the accident.

associations, the associations were struggling to grow first, before they could embark on facilitation of such services to the members.

Promotive measures

These are measures that aim to enhance the livelihood of workers through, for instance, income stabilization, thus including access to micro-credits/loans/saving and opportunities for training.

Informal construction workers associations both in Dar es Salaam and Dodoma have been trying to implement promotive measures in a small scale. As presented in Figure 8.2 below, when association members were asked what they considered to be the most important benefit of being a member of an association, 26.5 percent considered the opportunity to save and receive loans as the most important benefit of being a member of an association, while 18.1 percent mentioned social/financial support such as school fees, weddings, funerals, and sickness, and 10.8 percent each mentioned enhanced information on new jobs and higher income. The data indicates that quite some associations engage in savings, loans, and financial support and that this is considered key benefits by many. Since most of the construction associations are considerably young, some forms of promotive measures such as facilitating access to micro-credits were the future wishes of many of the associations.

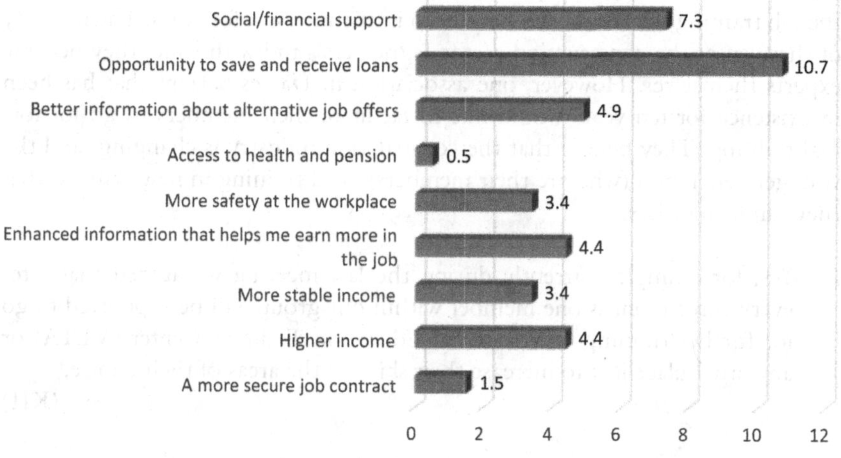

Figure 8.2 The most important benefit of being a member of an association.
Source: Author's elaboration based on project survey data.

Respondents from KIIs and FGDs considered the associations to be of benefit when in life emergencies. Most of the associations have saving mechanisms and provision of small loans with very minimal interest for the members. For example, when in a focus group discussion, one participant responded:

> This is very important for us, myself before coming here I came from collecting a TZS 25,000 from my association because I had a family emergency, and I will pay TZS 30,000 so the interest is TZS 5,000 only.
>
> (FGD participant)

For the informal construction workers, these associations act as a very fast and very reliable source through which one gets financial assistance urgently. It was also revealed from one strong group that workers could approach the group and take a loan to buy properties such as land, if a member is trustworthy, the group provides them with a loan, and they repay with interest as noted below:

> For example, the past six months one member asked for a loan to buy a piece of land, we sat and agreed on how he will repay and was given the money and he bought the land. That is in our constitution to help our members.
>
> (KII)

Since most of informal construction workers do not get jobs every day, as construction work is mostly irregular, the presence of associations makes their lives a bit more secure. Work in the construction sector does not necessarily require expert skills; thus, there are many people joining the industry to search for jobs. When these kinds of people join the association, they are informally trained (on the job training) by those who have been in the industry for years. The majority of these enter the construction sector as 'helpers', and with time, they become experts themselves. However, one association in Dar es Salaam that has been in existence for ten years would like to facilitate their members to go for formal training. They believe that the construction industry is changing, and the younger generation (who are their members) need training in new skills so that they can help others.

> Yes, for example, currently during the last meeting we agreed that after every three months one member within our group will be supported to go for further training at Vocational Education Training Center (VETA) or any other place just to increase their skills in the areas of their choice.
>
> (KII)

Transformative measures

According to the Devereux and Sabates-Wheeler (2004) framework, transformative measures seek to address concerns of social equity and exclusion, and this includes collective action for workers' rights among others. For the purpose of this research, transformative measures include voice, representation, and advocacy.

Discussions in FGDs that had a representation from many different associations as well as the KIIs showed some aspects of voice and representation. These transformative measures appear not to be the most important organizing factor, perhaps because the associations are young, sector-specific and mostly not registered. From KIIs and FGDs, we see association leaders representing their members in various meetings regarding tenders and other infrastructural development projects that are tendered by the local government leaders but few representations when a member faces a challenge like being in police custody or in employment negotiations. This is seen as a more personal issue that the associations should not be involved with. Association leaders agreed that it was only when the matter involved the association directly that it is taken on board. There have also been circumstances when the association gets a construction assignment and payments become a problem; in such cases, we see association leaders taking charge to follow up as narrated by one KII:

> There was a manager who gave us a job but at the end he didn't pay us when we completed the job, we then had to appoint representatives amongst us who went to the local government leaders to report so that we can be paid our money. It also happened once, a member of our group disappeared with our working tools, we also had to report to the authorities, and we got the support and our tools were returned. In all the situations, the authorities stood for us and we gained our rights.
>
> (KII)

Access to associations

Worker characteristics by association membership

Membership to associations is commonly open to only people within the sector, as confirmed by 76 percent of the respondents who noted that, for one to join the association, he/she must be a construction worker. Findings regarding various characteristics by association member status are presented in Table 8.3 below which shows the comparison between members of associations and non-members. Discussing the key characteristics such as formal social insurance enrolment reveals that there is no statistical significance between being a member or a non-member. Thus, although the incidence is higher for association members, membership does not seem to be a key factor for one to receive social insurance among ICWs. There was, however, a statistical significance in membership to association for those who lived in Dodoma as compared to Dar es Salaam, with only 33 percent of the members living in Dar es Salaam, compared with 67 percent earlier in Dodoma. The findings reflect the reality comparing the two locations. As seen in Table 8.1, most of the people living in Dar es Salaam have migrated to Dar es Salaam from other regions, whereas Dodoma is still a region dominated by locally born workers. Hence, it can be that being a local born in a somewhat semi-urban region means that people are more likely to join associations.

Table 8.3 Differences in key workers characteristics by association member status

	Member	*Not Member*	*Difference*	*t-Value*
Social protection enrolment	0.20	0.16	0.04	0.74
Health insurance coverage	0.22	0.14	0.08	1.45
Dodoma	0.67	0.48	0.19	2.74***
Dar	0.33	0.52	−0.19	−2.74***
Gender (male = 1)	0.98	1.00	−0.02	−1.73*
Married	0.67	0.69	−0.01	−0.21
Local	0.46	0.37	0.09	1.27
Age	39.07	34.55	4.52	3.46***
Mean daily earnings (current USD)*	11.22	17.17	-5.95	-1.89*
Assets (house and/or land)	0.60	0.43	0.18	2.50**
Primary incomplete	0.06	0.08	−0.02	−0.58
Primary complete	0.58	0.55	0.03	0.41
Secondary or above	0.36	0.37	−0.01	−0.11
Training course	0.31	0.22	0.09	1.48
Training on job	0.35	0.35	0.00	−0.04
Self-taught	0.34	0.43	−0.09	−1.28
Wage-worker	0.20	0.40	−0.20	−3.01***
Own-account	0.60	0.43	0.17	2.38**
Micro-business	0.19	0.16	0.03	0.53
Observations	205			

Source: Author's elaboration based on the project survey data.
Notes: <0.01***, <0.05**, <0.1*.

Membership to associations is also statistically significantly correlated with the age of the worker as members of associations have a slightly higher mean age (39 years) compared to those who are not members of associations (35 years). The results signify that, perhaps as one grows older, the need for associations becomes a crucial factor due to different factors such as family responsibilities, saving, and access to micro-loans which are not the most crucial needs for the younger people in the sector.

Furthermore, and interestingly, in terms of mean wage, those who are not members of association have a higher mean wage (USD 17) as compared to those who are members (USD 11). The reason could be that non-members are likely to be experienced workers working as gang leader *fundi mkuu* who have been in the job for long time or have established many connections with employers while association members are coming together forming association to search for jobs. This can also be explained by possession of assets, where members of associations are more likely to own a house and/or land compared to non-members. This is because members of associations have less worry on spending money on medical bills and other contingencies since they have opportunity for cushioning and loans.

Findings also shows that wage workers are more likely not to be members of associations, whilst own-account workers are more likely to be members, possibly related to the higher income earned by the latter compared with the former worker type. Although the distinction might not always be clear-cut, observation done by the researcher during the study noted that, for those who had worked in large construction sites (above 10 storey), the majority regarded themselves as wage-workers. While those members and non-members who have worked on medium construction sites (3 storey to 6) considered themselves to be own-account. The latter group of workers go to the construction sites with their own tools sometimes and thus feel that they are more own-account, offering their labour to the contractors and site engineers.

Entry conditions, engagement, and trust

As already explained, every association has different conditions to be fulfilled before being accepted as a member. For informal construction workers, the demonstration of commitment to membership and being active is through contribution to welfare funds; in some associations, this is independent from the monthly contributions. Some associations due to provision of micro-loans to members are constrained with resources when a member needs immediate financial assistance. In such cases, members are required to contribute from their pockets to subsidize the contribution that could be provided by the association.

Honesty and good character are also among the top qualities for being accepted to be a member because associations operate on goodwill. They are yet to have formal mechanisms of handling issues such as having formal bank accounts. In some groups, one member is trusted and elected to be a treasurer and thus entrusted with all group finances. Such members are trusted, and that is why, it is a requirement of all the associations that, before being invited, a member should be introduced by an older member who knows them well. Other characteristics such as participation in groups' activities as they arise, following the rules and regulations stipulated by the group or group's constitution were also important factors because members consider togetherness as the key strength of their associations. Hence, one exclusion criteria is deviation from rules as members can be expelled from the association if they are considered not serious.

Access to formal social insurance

Different institutional arrangements have been made available for workers in the informal sector to enrol in formal social insurance schemes voluntarily. However, it is evident that construction associations are not primarily geared towards issues regarding facilitating access to these schemes. As seen in Table 8.1, the survey indicated a very low enrolment rate to formal social insurances of only 18 percent, yet with some variation by location. When looking at which parameters differentiate SI contributors from non-contributors, Table 8.4 reveals that those receiving SI are more likely to be married, locally born, have higher earnings and assets, and be a micro-business. Married workers have more family responsibilities that might demand them to have a means of cushioning when faced by life challenges.

Table 8.4 Differences in key workers characteristics by social insurance enrolment

	Social Insurance	*No Social Insurance*	*Difference*	*t-Value*
Association member	0.46	0.39	0.07	0.74
Sacco/vicoba/chama	0.05	0.10	-0.04	-0.80
Work-related association	0.41	0.29	0.12	1.43
Women/youth/religious	0.00	0.01	-0.01	-0.66
Work-related	0.30	0.20	0.09	1.26
Loans	0.11	0.11	0.00	0.02
Voice and representation	0.00	0.00	0.00	
Social cushioning	0.05	0.08	-0.03	-0.60
Dodoma	0.65	0.54	0.11	1.19
Dar	0.35	0.46	-0.11	-1.19
Gender (male = 1)	1.00	0.99	0.01	0.66
Married	0.81	0.65	0.16	1.85*
Local	0.57	0.37	0.20	2.24**
Age	37.51	36.13	1.38	0.81
Mean daily earnings (current USD)	20.71	13.45	7.27	1.80*
Assets (house and/or land)	0.62	0.47	0.15	1.67*
Primary incomplete	0.00	0.09	-0.09	-1.90*
Primary complete	0.57	0.56	0.01	0.09
Secondary or above	0.43	0.35	0.08	0.93
Training course	0.27	0.26	0.01	0.18
Training on job	0.24	0.38	-0.13	-1.52
Self-taught	0.49	0.37	0.12	1.32
Wage-worker	0.32	0.32	0.00	0.03
Own-account	0.41	0.52	-0.12	-1.30
Micro-business	0.27	0.15	0.12	1.67*
Observations	205			

Source: Author's elaboration based on the project survey data.
Notes: <0.01***, <0.05**, <0.1*.

Due to the risky nature of construction activities, one would think that most workers would be tempted to enrol in formal insurance schemes, or the association could facilitate the move to join the schemes. When asked why they do not contribute to an insurance scheme, at least 25.4 percent said that it was too expensive, while 20.5 percent said that they had no knowledge of insurance. Others noted that the services were poor (11.7 percent), while about 9.3 percent said that they have never given National Health Insurance Fund (NHIF) registration attention. There was also a small percentage (2.9 percent) of people who preferred to save by themselves, among other reasons, as reflected in Figure 8.3.

However, findings from KIs and FGDs suggest that, although workers are aware of the presence of these social insurance schemes and consider it to be important, most respondents point to the issue of irregular incomes and how to manage the same as a key hindrance. They also noted financial management illiteracy as one of the reasons for not contributing to these schemes. They therefore

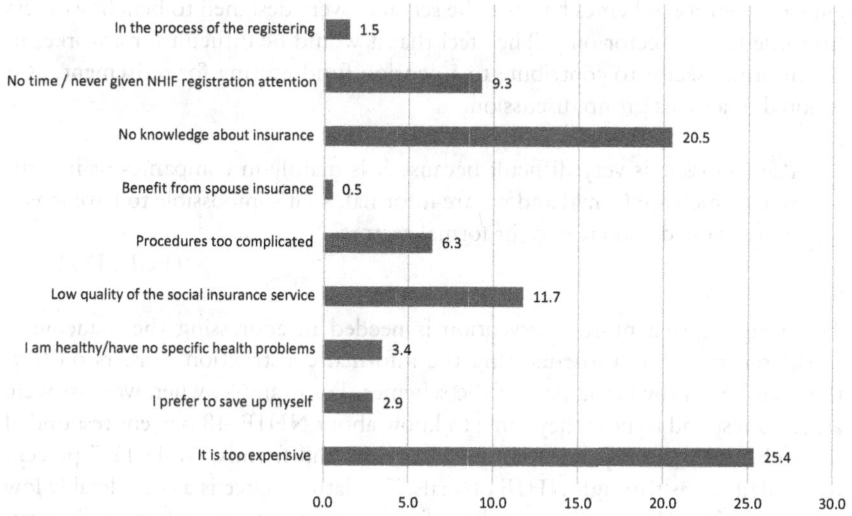

Figure 8.3 Reasons for not contributing to an insurance.
Source: Author's elaboration based on project survey data.

have mixed feelings as to why it is important to enrol in an insurance scheme as noted in in one FGD:

> ...The issue of insurance especially health insurance is very important, but we have not been educated on its importance. I think we need more education on how we can manage our little incomes and be able to contribute. I think every 'fundi' can manage but there is little awareness of its importance.
> (FGD, Dar es Salaam)

The FGD revealed the high premium costs as a key reason why workers do not enrol. Due to the irregularity of construction activities, workers are not sure on the availability of jobs and therefore scared to commit resources to insurance contribution as highlighted below:

> For example, our incomes are not secured, you can have money today but stay for a month without a job, how do we manage this...
> (FGD, Dar es Salaam)

However, there is also among some a very positive view on the importance of insurance, especially health insurance which is seen as very important. Workers acknowledged that when one has a health insurance for themselves and their families, people are able to manage their incomes to run other family needs instead of spending it in hospitals.

There is a widespread feeling that informal construction workers cannot benefit from pension schemes because the schemes were designed to benefit workers from the formal sector only. They feel that it would be difficult for a worker in the informal sector to contribute to a pension fund waiting for retirement, as it is noted in a focus group discussion.

> Pension issue is very difficult because it is mainly in companies or institutions which are formal and we are informal, so it's impossible to have it as it was designed and created for formal sectors.
>
> (FGD, Dodoma)

This suggests that more intervention is needed in addressing the inadequacy of the schemes, but also educating the informal construction workers on how they can benefit by being part of the schemes. For example, when workers were asked to respond to how they came to know about NHIF, 48 percent responded through media, 19.5 percent through relatives and friends, while 12.7 percent responded it was through NHIF officials. The latter source is a considerably low percentile since the locations of the informal construction workers are known, and the officials could locate them easily or organize a workshop and invite them.

Conclusion

Formal social insurance interventions are an important factor that enables individuals to deal with life vulnerabilities around the world. Evidence from this study has revealed that informal construction workers in Tanzania are lagging behind in accessing formal social insurance measures. There is a need for more comprehensive and inclusive mechanisms of enrolling ICWs in insurance schemes in order to address the challenges they face. Findings from this study have shown a considerable level of awareness of the presence of formal social insurance, which could be a significant entry in designing a more focused scheme that allows workers with unpredictable and irregular incomes to enrol and benefit. Additionally, it could be a point for the officials to provide critical education on the benefits of social protection measures.

Informal construction workers seem to have an alternative to formal social insurance interventions. Some of the associations provide a strong cushioning mechanism which is flexible and seems to fit well with workers' needs. This could explain, to some extent, the low enrolment into formal schemes. As noted, worker associations appreciate the availability of formal social insurances such as the health insurance and pension. However, there is a need for capacity building to association leaders to be able to formally register the associations, link up with the formal systems, and create favourable mechanisms that would suit the needs of the informal construction workers and their associations.

Informal construction workers are not able to demand rights when working on construction sites. Partly because of the large pool of people with the same skills – thus, if you are not able to work, you can easily be replaced. Serious injuries and other occupational accidents have been prevalent in the sector, but still findings show that it is rare that the contractor, client, or site engineer takes full responsibility of the injured person. There has been evidence where the site engineer/owner/contractor have taken the initial steps such as first aid and initial hospital costs, but, thereafter, it is relatives, friends, and associations that take the responsibility. The lack of contracts and formal agreements to hire labour is also a contributory factor. Due to lack of contracts, liability is to a great extent removed from the hands of the so-called employers.

Associations are generally very informal, not registered, and thus lack the power to represent their members fully. They are organized mostly to deal with welfare issues such as occupational injuries and accidents, funerals, and assisting members on other social welfare issues as well as getting jobs. They therefore lack the potential power to influence decisions that would benefit or otherwise affect them such as being able to have a representation in the decision-making bodies. Moreover, the trade union (TAMICO) has not prioritized the issues of informal construction workers and their associations.

The government of Tanzania, through the current pension fund (NSSF), has established a voluntary scheme that members of the informal sector, including construction workers, could join, but workers in the informal sector are not enrolling in the scheme. The current study revealed that access to formal SI schemes is tedious to members. The lump-sum premium is expensive, and workers have no education on what the benefits are; however, there was indication that if given education they could join. An easier set mechanism could assist in the promotion of awareness, registration, and follow up of how workers in the construction sector could directly benefit through the established voluntary schemes.

Notes

1 For more detail on the overall methodology, refer to Chapter 1 and Chapter 3 appendix.
2 There were no variations in terms of association services and activities across Dar es salaam and Dodoma.
3 Based on interview with the Chairman.
4 Based on interview with the Chairman.

References

Devereux, S. & Sabates-Wheeler, R. (2004) *Transformative social protection.* IDS Working Paper 232. Brighton, Institute of Development Studies.

ILO (2017) World social protection report 2017–19: *Universal social protection to achieve the sustainable development goals.* International Labour Organization. Geneva, International Labour Office.

Jason, (2007) *Informal construction workers in Dar es Salaam, Tanzania.* Working Paper 226, Sectoral Activities Programme, International Labour Organization. Geneva, International Labour Office.

Mkenda, B. & Aikaeli, J. (2015) *Informal construction employment, earnings and activities: A boon or bane for Tanzania?* Available from: doi:10.2139/ssrn.2706036.

Mrema, E.J., Ngowi, A.V. & Mamuya, S.H. (2015) Status of occupational health and safety and related challenges in expanding economy of Tanzania. *Annals of Global Health.* 81 (4), 538–547.

NBS (2013) *Unemployment estimates for the year 2011.* National Bureau of Statistics. Dar es Salaam. Available from: http://www.nbs.go.tz [Accessed 25 February 2017].

NBS (2019) *Tanzania in figures.* Dar es Salaam, National Bureau of Statistics.

Schmalz, S., Ludwig, C. & Webster, E. (2018) The power resources approach: Developments and challenges. *Global Labour Journal.* 9 (2), 113–134.

Silver, B.J. (2003) *Forces of labor: Workers' movements and globalization since 1870.* Cambridge, Cambridge University Press.

Wells, J. & Jason, A. (2010) Employment relationships and organizing strategies in the informal construction sector. *African Studies Quarterly.* 11 (2&3), 107–123.

World Bank. (2018) The World Bank annual report 2018. Washington, DC, World Bank. © World Bank.

Wright, E.O. (2000) Working-class power, capitalist-class interests and class compromise. *American Journal of Sociology.* 105 (4), 957–1002.

9 Construction workers in Kenya

Straddling with formal and informal social protection models

Winnie Mitullah

Introduction

The construction industry in Kenya is multi-layered with most workers working under sub-contractors, contractors, and clients, who are also mostly informal and not registered with authorities. This makes it difficult to monitor and enforce labour laws, regulations, and standards in the sector. The workers have neither an employer who can share the cost burden that Social protection measures entail nor do they have rights as required by government policies, regulations, and standards. Most workers are hired on a short-term basis ranging from daily, a few days, weekly, and monthly with very few having steady employment with one employer for more than six months. This makes them not qualify for any worker benefits since they are considered casual workers. In a few cases, employers cater for some aspects of their social protection but without full commitment, leaving workers straddling between formal and informal social protection models.

Chapter 4 (Bill of Rights) of the Constitution of Kenya (ROK, 2010) guarantees all Kenyans' social, economic, and cultural rights. It calls upon the state to provide appropriate social security to persons unable to support themselves and their dependents which is closely linked to other rights including the right to health, human dignity, reasonable working conditions, and access to justice. The Kenya Vision 2030 complements the Constitution with its three key pillars of economic, social, and political aspects (ROK, 2007). The objective of the social pillar is investing in the people of Kenya in order to improve the quality of life for all. A goal guided by the National Social Protection (NSP) policy, which identifies three aspects of social protection which are geared towards improving human capability, that is, social assistance (e.g. cash transfer programmes), and social security consisting of pensions and health insurance. These three social protection measures fall under what in the Devereux and Sabates-Wheeler (2004) framework is called preventive and protective measures. In addition, this framework also includes promotive and transformative measures as discussed in Chapter 1 of this edited volume.

Social protection is a safety net mechanism which is crucial in building the human, financial, natural, and physical assets of poor people enabling them to grow out of poverty (Shepherd, Marcus & Barrientos, 2004). The informal

DOI: 10.4324/9781003173694-9

construction sector has many poor workers who are not entitled to most labour rights provided by Kenya laws and regulations such as the Kenya Occupational Safety and Health Act (OSHA) of 2007 and the Work Injury Benefit Act (WIBA) of 2007. The latter provides for compensation to employees for work-related injuries and diseases contracted in the course of employment. Under WIBA, every employer is expected to compensate employees for any death or injury while on duty and cater for medical expenses and funeral costs. WIBA only insures permanent workers; yet, over 80 percent of construction workers are informal and are therefore not entitled to insurance.

Labour regulations, including Occupational Health and Safety (OHS) rules governing construction workers, have increasingly been flouted in many parts of the world, as formal enterprises employ casual labour or outsource their labour in order to avoid the costs associated with employment (Mitullah & Wachira, 2003; Wells, 2007). The Indian Code of Social Security, 2020 provides a 2-percent construction levy collected by Local Authorities and credited to the Building and other Construction Workers Welfare Fund (COSS, 2020). This is aimed at cushioning any construction worker who works continually for three months so long as they are registered with the Construction Workers Welfare Board provided by the Code. By levying a fee on construction industry, social protection schemes can be designed to reach specific occupational groups in the informal sector. While these are good proposals, realizing them is a challenge considering that many workers are informally employed by informal labour contractors and have weak agency and organizational power, as discussed in this chapter. The Indian Code acknowledges this and provides for portability across states, those who work continually for three months, and social protection after leaving work so long as one has been in employment for three years.

The chapter explores how construction workers are organized, how worker associations contribute to labour agency and social protection, and the role played by formal institutions and institutional structures in strengthening or undermining workers organizing. First, the chapter outlines the characteristics of construction workers and examines the associational landscape using selected associations. An analysis of how the power resource approach (PRA; Schmalz, Ludwig & Webster, 2018) applies to the sector and how preventive, promotive, and transformative services are offered by workers' associations follows. These subsections set the ground for a discussion on access to associations highlighting entry requirements, comparison of association member status, access to social protection, and finally, a conclusion.

The chapter is informed by data drawn from a larger study of informal economy worker organizations and social protection covering Kenya and Tanzania (see Chapter 1 of this edited volume for details). In this chapter, the cities of Nairobi and Kisumu in Kenya are used as case studies. A combination of quantitative and qualitative research methods was used to gather information from construction workers and their associations. A partly random survey of 221 construction workers in the two sites was carried out between June and December 2018. This included workers sampled using geographical location

Table 9.1 Key worker characteristics

	All		*Nairobi*		*Kisumu*	
	Mean	*SD*	*Mean*	*SD*	*Mean*	*SD*
Member*	0.33	0.47	0.33	0.47	0.33	0.47
Formal social insurance enrolment (health/pension)	0.30	0.46	0.33	0.47	0.26	0.44
Health Insurance coverage	0.35	0.48	0.37	0.48	0.34	0.48
Gender (male = 1)	0.90	0.29	0.91	0.29	0.90	0.30
Age	35.77	9.42	35.89	10.20	35.63	8.39
Married	0.79	0.41	0.78	0.42	0.80	0.41
Local born	0.22	0.42	0.11	0.31	0.37	0.48
Mean daily earnings (current USD)**	10.81	8.70	11.70	8.27	9.70	9.13
Assets (house and/or land)	0.29	0.46	0.31	0.46	0.28	0.45
Primary incomplete	0.07	0.26	0.07	0.25	0.08	0.28
Primary complete	0.39	0.49	0.41	0.49	0.37	0.48
Secondary and above	0.54	0.50	0.53	0.50	0.55	0.50
Professional training course	0.35	0.48	0.34	0.48	0.36	0.48
Training on job	0.48	0.50	0.48	0.50	0.48	0.50
Self-taught	0.17	0.38	0.18	0.38	0.16	0.37
Wage-worker	0.95	0.23	0.93	0.25	0.96	0.20
Own-account	0.03	0.16	0.03	0.18	0.02	0.14
Micro-business	0.03	0.16	0.03	0.18	0.02	0.14
Observations	221		123		98	

Source: Author's elaboration based on the project survey data.
Notes: * Based on the random sample consisting of 170 workers. ** The median wage is USD 9.4 cent overall, USD 9.9 for Nairobi and USD 8.7 for Kisumu.

(75 percent) and membership in associations (25 percent). In Nairobi, a geographical (random) sample of 94 workers and 29 from associations yielded a total sample of 123 workers (55.7 percent). In Kisumu, a geographical (random) sample of 76 and association membership of 22 provided a total sample of 98 (44.3 percent) workers (Table 9.1). In order to achieve the geographical sample, constituencies and wards in each of the two cities were mapped to identify the ongoing construction work in commercial buildings. In addition, Focus Group Discussions (FGDs) and Key Informant Interviews (KII) were used to gather information.

Characteristics of construction workers

The summary of key worker statistics (Table 9.1) reveals that the construction sector is dominated by married (79 percent), male (90 percent), and wage workers (95 percent) with a mean age of 36 years. The majority of workers had completed secondary education, had different forms of training which included a high percentage of training on the job (48 percent), professional training (28 percent), and self-taught (17 percent). The sector is not strong in associational life based on random sample showing 33 percent compared to trade and transport sector which had 47 percent and 57 percent, respectively.

Around a third (30 percent) are enrolled in formal social insurance (SI), which could be the National Hospital Insurance Fund (NHIF) or the National Social Security Fund (NSSF) or both, with the majority (18 percent) referring to the former. The median wage is USD 9.4[1] which is lower than the mean due to income disparities among construction workers with a few heavy earners at the top pulling up the mean. The income of those who are members of associations is higher (Table 9.2) which partly demonstrates that membership to associations is beneficial. Associations assist members to find jobs, cushion members during difficult times such as bereavement, and also give loans, some of which are used for boosting income. These issues were largely similar in Kisumu and Nairobi except for SI enrolment which is higher in Nairobi and local born which is higher in Kisumu, while wage is lower. Most Kisumu workers are born within the region and are likely to have additional social cushioning from their communities, a provision which is not available to Nairobi construction workers who come from across Kenya.

Working in the construction sector has challenges and threats, which are largely related to informality manifested in the lack of legal contracts. Survey responses on whether workers prefer contracts specifying pay and benefits yielded mixed responses, with many noting that availability of contracts would protect them from non-payment of dues, while a few others prefer being their own bosses with daily payment. Overall, the majority want job security and assurance of benefits, while others prefer working without a contract for flexibility. These findings were further corroborated in FGDs with some workers noting that, if given a chance, they would advance their work, earn more income, and buy assets.

Associational landscape

This sub-section provides an overview of construction workers' associations including a brief on selected associations highlighting the different forms of associations, how associations are organized, and the services provided to members. This is followed by reflections on how the PRA applies to the construction sector.

Overview of associations

Construction associations provide several benefits to members of which opportunity to save and receive loans (39 percent) and work-related (50 percent), including information on employment opportunities (31 percent) (Table 9.2), are the main attractions for joining associations. They play a central role in providing opportunity to save and receive loans, linking members to employment opportunities and contacts for jobs as elaborated by a key informant from the Labour Link association (Box 9.1).

Other benefits of being a member of a worker association such as safety at workplace, stable income, and protection against harassment are secondary. The

Box 9.1 Associations as a link to employment

We can be contracted by clients as far as Nairobi. I always try to convince the client that I am a painter. We do not strain ourselves looking for clients. Within the Labour Link Network, we have worked for big hotels such as Acacia, United Mall, and Sunset. When our clients have work, I am the one who does it ... we have bought tools, we do not hire tools anymore. We sit down and agree what percentage of the money we receive from contracts done will go to buying tools – we are like a family, we support one another. Once we get a contract, we get our people, and if they are not enough, we hire others in the places where we have gotten the job (KII, Kisumu).

findings reveal that associations are very good at supporting members in getting jobs but generally weak in aspects such as ensuring secure contracts, access to health and pension, and higher incomes. As seen earlier, based on the random sample, only 33 percent of construction workers belong to associations, which are largely work-related associations.

An examination of workers who are association members revealed that 53, 40, and 7 percent, respectively, belonged to work-related associations, SACCOs, and other associations (Table 9.2). Thirty-six percent were registered in SI schemes showing a higher incidence of SI among members of associations

Table 9.2 Key worker characteristics, association members

	All		*Nairobi*		*Kisumu*	
	Mean	*SD*	*Mean*	*SD*	*Mean*	*SD*
Formal social insurance enrolment (health/pension)	0.36	0.48	0.37	0.49	0.34	0.48
Health insurance coverage	0.39	0.49	0.40	0.49	0.38	0.49
Association type						
Sacco/vicoba/chama	0.40	0.49	0.38	0.49	0.43	0.50
Work-related association	0.53	0.50	0.55	0.50	0.51	0.51
Other association	0.07	0.25	0.07	0.25	0.06	0.25
Benefit type						
Work-related	0.50	0.50	0.63	0.49	0.32	0.47
Social cushioning	0.10	0.31	0.00	0.00	0.23	0.43
Voice and representation	0.01	0.10	0.02	0.13	0.00	0.00
Loans	0.39	0.49	0.35	0.48	0.45	0.50
Barriers	0.64	0.48	0.60	0.49	0.70	0.46
Fee	0.73	0.45	0.58	0.50	0.91	0.28
Observations	107		60		47	

Source: Author's elaboration based on the project survey data.

compared with the full sample of construction workers. The association members had a health insurance coverage of (39 percent), with a standard deviation of 0.49. Voice and representation which falls under transformative social protection was hardly visible in the survey and hardly mentioned in FGDs and KII discussions.

Highlights of selected associations

Associations in the construction sector take different forms which include registered associations, loose[2] unregistered networks, cooperative societies, and companies. These associations have rules for entry including subscription requirement and support construction workers in looking for employment opportunities and welfare needs as discussed below.

Labour Link, established in 2012, is a loose group based in Kisumu. It is composed of five core leaders, each with specific construction sector skill sets who rally individuals with similar skill sets to form work cohorts. Formation of these cohorts enables the group to easily interact, call each other when there is work, and compete for jobs. The group has 20 members who take piecework assignments moving from one assignment to another, and in cases of more jobs, they invite others. The five leaders come from one village, but those operating under their clusters do not necessarily come from their respective villages.

The network has assisted members to buy tools and develop their skills, as noted by one of the group leaders, "we can do paper walling in a professional manner and we get called for jobs in big commercial projects". They listed several well-known commercial buildings in Kisumu, in which they have been employed to handle different aspects such as painting and fittings. Whenever they get a contract, they agree on the percentage to be deducted and invested in buying tools for the group.

Despite the activeness of the group, it is not registered, and there is no membership fee. There was an attempt to register the group, but it did not work, largely because of loan repayment defaulters. The group was running a SACCO with members taking loans and not paying back. The poor experience with loan repayment resulted in the decision of the group to remain informal and only assist each other in social needs and in getting jobs.

Migosi Friendly Workers Group in Kisumu, established in 1994, has an investment arm called Migosi Builders, established in 2008. In both FGDs and in KII with the chairman, it was noted that the association is very useful in assisting members to get jobs, certification, and negotiation with employers. The friendly group has 140 members, while the builders have 42 members, 4 are women. The workers' group which is largely composed of masons and three certified electricians is registered with the Department for Social Services. The membership has kept increasing since it is difficult to get a job in the construction sector when one is not in a group. The group is well organized with clear procedures of rotational recruitment and governance, although the incumbent leaders had been re-elected four times.

The group enrols new members including standard eight primary school drop-outs. They join as helpers and progress to become masons. Members pay a registration of KES 200, and once registered, members pay weekly share contributions of KES 500, while those in the investment arm register with KES 2000 and pay monthly fees of KES 800. Membership is open to any construction worker referred by a member. A copy of national identification card, physical contact, and name of area chief is mandatory. It was emphasized by an official of the association that these requirements are necessary as a point of reference since not all individuals have a good reputation. The association provides a contact point for professional construction workers, upcoming workers, and young school leavers willing to work in the industry. The association also gives workers opportunity to operate, assists members to acquire certification, in negotiation, including intervening when disagreement occurs or contracts are breached by clients. In cases where negotiation has failed, one is free to go to the government authorities.

The association applies group dynamics to access rights. An example was given in which group members teamed up to demand payment for unpaid dues of workers, to quote a respondent, "one day, I found a gang baying for the client's blood with *pangas* (machetes) because some team members had not been paid". A probe into payment modes and challenges revealed that, in some cases, contractors are not paid by their clients, in particular, government who takes long to pay. In such cases, associations intervene through the leadership, which encourage members of the group to report to the leadership whenever there are conflicts, including non-payment of dues, instead of taking punitive measures on the contractors and clients.

Chama Cha Mafundi is a semi-structured non-registered loose network of 68 members established in 2003 whose contact (waiting point) designated location is on a street in Nairobi – Moi Avenue within the Central Business District (CBD). At the time of research, only 20 members were active. The group reports to the open space every day, and clients come and procure their labour. The network has a constitution specifying rules and regulations for members. Each member deposits KES 200 (USD 2) to the association whenever they secure a job. Group members know each other well, and each person is obliged to be honest and contribute to the group kitty which is mainly used for welfare. Members are free to borrow from the kitty in case of serious needs.

Membership fee is KES 1,000 (USD 10), and new members must be introduced, and if they do not have the amount, the one introducing them can pay on their behalf. Membership is open to all in the construction sector. The association puts emphasis on integrity and does not entertain dishonesty in construction sites or off-site. An official of the association noted that this is stipulated in the constitution which guides members on their daily operations. Members engage in building, renovation, and repair work and are willing to take jobs anywhere within and outside the city of Nairobi.

There are many benefits of being a member, with welfare being the major one. They support each other in sickness and undertake fund-raising for school fees and to pay fines when a member is arrested for what they consider to be unavoidable such as working on unlicensed sites. They also advise members to ensure that clients have a license for construction, especially within the CBD and commercial areas to avoid problems with the city authority. However, the group seemed more relaxed with individual clients as opposed to corporate clients. The group also provides a link to employers and advises members on what to demand from clients, for example, toilet facilities if work lasts more than three weeks.

The association was previously giving loans to members, but the activity collapsed when the group had raised capital amounting to KES 680,000. Some members borrowed but never paid. In 2018, the group decided to stop the savings scheme and share the money among members in line with their shares. The group was considering rallying members once again to get back to saving, but were concerned about some conservative members, especially old guards – largely painters who were the founders of the association. They were noted to lack innovation and not interested in advancing associational activities such as development of SACCOs.

The group members are supportive of each other, occasionally assist members to negotiate with clients. Their meetings every two weeks revolve around issues of relationship with their clients and employers. Welfare is hardly discussed but handled on case-by-case basis, and members are largely ignorant of the importance of associations, to quote, "members want jobs and that is all, and not discussion of associational issues". This had also contributed to poor attendance of association meetings, which hardly attract half of the membership, and the association has not recorded any significant changes in 14 years of operation. The meeting and recruitment base have been operational for over 30 years and had the potential of developing a strong network of construction workers, but remain fragile and not effective in addressing labour rights issues.

Three active members of Chama Cha Mafundi had established and registered a company with the National Construction Authority (NCA). The three decided to go a notch higher to promote their construction work. The establishment of the company was driven by the challenges of getting jobs as individuals. One of the proprietors had left the association for work with a company, came back, re-joined the group, and established a company. This shows the synergy between informal and formal work and the influence interaction with one sector has on the other. However, this synergy has not directly benefited the association but only a few individuals. This could probably be explained by the laid-back nature of the association and its inability to advance the interests of members. This notwithstanding, entrepreneurial individuals can leverage associational platforms to advance their opportunities as done by these three members.

FundiTech Services Cooperative Society in Nairobi was registered in 2017 with a vision of being a leading workers' cooperative for professional technical services in Kenya. The cooperative originally began in 2007 as a welfare group under the name Mafundi wa Kenya (MWAKA) association with a main goal of providing work-related information for *fundis* (masons) and technicians. Many approached the association for information but were not willing to join the group. This triggered the group to think of how to attract membership by undertaking activities that benefit the members. A meeting with USAID advised the group to convert the association into a cooperative, opening the pathway for the group.

The cooperative has a total of 107 members, of which 15 are women, about 30 percent of these members are formally trained while 70 percent are not formally trained. Members have varied skills, including masonry, electrical, welding and fabrication, tile fixing, steel fixing, plumbing, and painting. Discussions with the secretariat of the cooperative revealed that skills are wanting even for those who are formally trained. Membership requirements include being a technician and a certificate holder who is trained with skills and specialization in the building and construction industry. Members should be 18 years and above, not in any other cooperative society carrying the same activity and be able to pay registration fees of KES 3,000 (USD 30) and a minimum share capital after a three-month probation period. Forty members have bought shares, of which three are women. The probation is a learning period which also gives the cooperative society a chance to know the member. Members are ranked as bronze, silver, and gold. These ranks are skill-based and do not depend on share-holding. The cooperative has a secretariat that coordinates all the services provided.

The cooperative has several benefits and obligations to members, a major one being growing a team of qualified construction sector workers, training members, and producing quality work. It is a labour sub-contractor with FundiTech getting contracts and doing supervision and management of labour while members provide labour. The cooperative has designed a formula of compensating members and ensuring quality work, with no *fundi* being paid less than KES 1,000 (USD 10) per day. The association has a 'daily must', specifying amount of work each skill must produce, for example, how much painting or building must be done in a day. Although the cooperative collaborates with other stakeholders majorly for advocacy, they have prioritized institution building, to quote an official of the cooperative, "we are not networking, we have discouraged people from coming to us ... construction is very dynamic, the *fundi* want to hear one word, their money".

The overview of the above construction workers' associations reveals that associations are diverse in structure and with different strengths, although assisting members to find jobs and cushioning members during sickness, bereavement, and other welfare-related issues cuts across all associations. The established and structured associations such as Migosi builders and FundiTech Cooperative perform better in supporting members compared to associations with loose networks such as Labour Link and Chama Cha Mafundi. While the workers tend to have foresight on their ability to work and improve their status, including collectively buying equipment as was noted by a Key Informant from Labour Link (see Box 9.1), they tend to have weak organizations which undermine their agency in voice and representation.

Associations such as Chama Cha Mafundi have operated for many years and remained a loose organization with a collapsed SACCO due to poor commitment of members. Members compete between themselves and lack trust, largely due to scarcity of employment opportunities, which contributes to members of associations flouting labour fees set by some associations. An interview with Migosi Builders association elaborated this point noting that, "somebody will bid very low to land a job, while some members hide to take such jobs due to desperation". The mistrust is intensified by failure by many members to pay their association's subscriptions and loans. Associations such as Labour Link resolved not to pool resources in the form of savings due to lack of repayment of loans. The SACCO of the Chama Cha Mafundi association also collapsed due to non-payment of loans.

Several of the activities being undertaken on behalf of the membership of associations can be understood using a PRA advanced by Wright (2000), Silver (2003) and other authors as discussed in Chapter 1 of this edited volume. The sub-section below assesses the applicability of the PRA on the construction associations.

The power resources of associations

The PRA which includes structural, associational, institutional, and societal power with its coalition and discursive tenets can be used to analyze operations of construction workers, since one third of the workers were members of associations. Authors (Wright, 2000; Silver, 2003; Schmalz, Ludwig & Webster, 2018) who use the PRA argue that workers can collectively mobilize different power resources to advance their interests and concerns. Analysis of construction workers' associations reveals that the workers mainly use associations to search for jobs, negotiate for terms of employment, and to informally intervene when disagreements occur between the workers and employers. This role of associations somehow mirrors what unions do in a formal context using a combination of associational and institutional power.

The findings reveal that a higher number of those who belong to associations are registered in SI schemes (see Table 9.3) and are able to secure their health and pension which are pillars for empowerment. However, the associational power of associations is weakened due to challenges such as members not paying association dues and participating in scheduled meetings. The inability to attend meetings and pay dues can be explained partly by governance challenges and the nature of construction work which involve long hours of work and movement to different sites, including relocating for some period for work outside cities of residence. This undermines associations' potential to leverage all associational power resources and is partly explained by the non-transparent, informal, and casual nature of the employment relations.

The construction associations are not able to use institutional power, although they are aware of the laws, regulations, and procedures, largely due to lack of formal contracts, time, and resources involved in legal processes, and for fear of losing employment opportunities. Associations such as Migosi Friendly Workers occasionally use associational power to team up and apply disruptive power (a form of structural workplace bargaining power) – storming work sites to demand payment from clients and contractors. Such storming would not be as necessary if the association had institutional power and operated as registered companies with legal contracts. Findings revealed that the contractors are hesitant to take legal action when the client is the national or county government when payments are not honoured. This is due to fear that the system might not favour them or that they might be blacklisted from getting future contracts/jobs. The fact that contractors also compete for jobs individually also weakens their ability for collective negotiation and redress.

The contractors and sub-contractors who employ workers operate in a quasi-formal manner having formal contracts with clients but hiring workers informally without contracts. Due to the weak structural and institutional power of the workers' associations, they are not able to argue their case and demand their rights from the contractors and subcontractors who employ them informally. In a few cases where workers' associations are collaborating with NGOs and other private firms, there are signs of societal power through coalition building and discursive power beginning to manifest.

Societal power was more visible in Fundi Online Platform which has nurtured relations with a university, a technical training institution, local companies, and an international NGO to realize its goal of being an online platform for workers. The Cooperative Society had also been assisted by USAID to advance its mandate, and a sign of agency and tapping of different sources of power were latent. The cooperative is ensuring compliance with laws and regulations and is able to protect the workers, albeit with financial and membership commitment challenges. The cooperative had prioritized institution building and was, in the meantime, not interested in nurturing too many external networks – which seemed a good strategy for consolidating the cooperative before venturing into the outside world.

There was no evidence of workers' associations nurturing relations with people in positions of power such as elected leaders, government, chamber of commerce, or trade unions to address issues facing them. The only link was through training offered by government agencies such as National Industrial Training Authority (NITA), NCA, National Youth Service (NYS), and private cement and paint firms. Some of the trainings were supported by private companies and facilitated by NCA as revealed in discussions with Kaberia Jua Kali Association which benefited from training by a Qatar-based company. The workers are largely inward-looking with many belonging to loose unregistered horizontal networks which connect them to employment opportunities. Associations such as Labour Link, Chama Cha Mafundi, and others that participated in FGDs belong to this group. Labour Link was investing in buying tools but was very poor in associational life with hardly any meetings taking place.

The few associations organized as cooperatives and companies engage all forms of power except disruptive power. A small company, an outshoot of Chama Cha Mafundi, has managed to register with NCA and to collaborate with formal construction industries and workers. The leader of this company had a short stint in the formal sector as a member of the group. This is a good demonstration of how institutional power assisted the construction worker to turn into an entrepreneur owning a company and still belonging to an association which had largely remained static. Such power tends to be limited among construction workers, and the question remains why the progressive entrepreneur was not able to influence the association to tap into institutional power such as registration with NCA and interaction with formal construction firms to advance associational power and workers status.

Services offered by construction associations

Construction associations offer mainly two services, informal social protection as well as assisting in search and link to employment opportunities. The social protection services offered to members can be assessed using Devereux and Sabates-Wheeler's (2004) classification of social protection measures grouped into four categories: Preventive, Protective, Promotive, and Transformative, three of which are discussed below.

Preventive

Legal provisions in Kenya require employers to ensure OHS of workers, subscription to the NHIF, and NSSF. However, findings reveal that the majority (70 percent) of construction workers do not subscribe to any SI schemes. The workers largely rely on informal SI strategies accessed through associations, family, relatives, friends, and good Samaritans. Association members collectively support their members in cases of accidents on-site as well as health issues and death in their homes. Some associations limit support for sickness to in-patient with cover limited to a member, their spouse, children, and parents. However, dependent beneficiaries only get about half the cover, while death of a member is treated in a limited manner, which requires additional fund-raising for attending funerals and taking extra responsibilities such as buying a coffin and transporting the body.

Associations also have ad-hoc informal SI support, in which there is no specified amount, but decisions are made by members when a member is confronted by a social issue which the group considers to require support. These range from sickness, death, school fees, weddings to paying local authority and government fines. The study revealed that even loose workers' associations like Chama Cha Mafundi are very good at rallying members to undertake fund-raising for sickness, school fees, and to pay fines.

Labour rights, such as safety at work in the sector, are handled in a multilayered manner and are not solely the responsibility of the employer. In cases of injury, medical expenses are paid either by the employer and workers, with each scoring 38 percent in the survey. Other cases are handled by the site manager or owner of construction, relatives, and work colleagues. This implies that workers through their associations are largely paying for medical expenses with the support of others. A further probe in FGDs and with KIIs revealed that

Box 9.2 Perception on safety of workers

Safety is more personal ... the worker must ensure his or her safety at work. Many contractors do not care about safety. Sometimes, the client forces contractors to enforce safety. In one incident, I was hanging on the fourth floor, and the client forced the contractor to provide a scaffold before anything could proceed. Safety is not a priority to clients, and workers cannot insist on safety gears, and if they do, they lose the job ... some contractors tell workers who ask for safety gears, 'choose whether you want a job or gears' (KII, Sub-contractor, Kisumu).

workers and associations do not use institutional power to address labour issues, but rely on associational collective power to support members. They view safety and accidents largely as bad luck and workers' personal issues as opposed to an employment problem as observed by a key informant in Kisumu (Box 9.2).

The voice in Box 9.2 reveals powerlessness and inability to address labour contractual issues which should be covered by WIBA. Instead, associations and social networks fill such deficits as noted by a sub-contractor "if they do not report to work, they do not get pay, if you are sick today, you will have no pay, but if you get better you will get back to work" (KII: Sub-contractor, Kisumu). Further discussions in FGDs and KIIs highlighted the support injured members get from associations and the need for members to take care of their safety as noted by the KII sub-contractor, "it is a must to contribute". The KII noted that he was responsible for the workers, and if injured, he ensures that they are taken to hospital. He further observed that the client is not responsible for those who get injured; in most cases, the clients are not present when accidents occur. Overall, the WIBA provisions were minimally applied, although most workers and employers knew the provisions.

Promotive

Construction associations are anchored on providing promotive services such as micro-credit, loans, savings, and training which they sometimes do in collaboration with other partners and agencies. As reflected in Figure 9.1, they put a high premium on savings and receiving loans, which are promotive aspects of social protection. Considering the precarious nature of their work, savings and loans are what cushion them by supplementing their irregular incomes and sustaining their livelihoods.

In the survey sample, the opportunity to save and receive loans (39 percent) and information on employment opportunities (31 percent) were the most important benefits of being member of a worker association, as reflected in Figure 9.1.

Micro-credit leveraged through associations sustains most of the workers, especially during periods when they are not able to get jobs. For example, Chama Cha

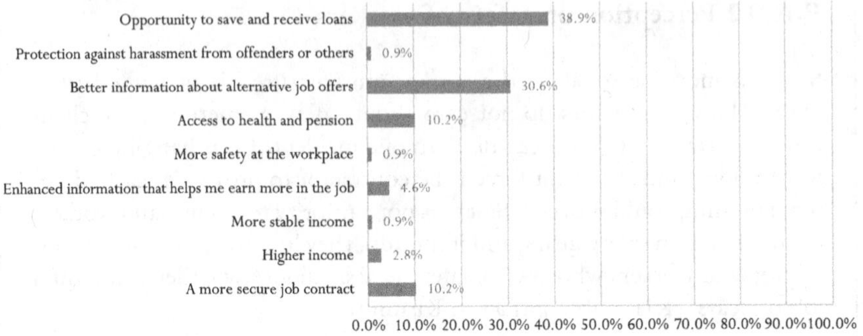

Figure 9.1 Most important benefits of being a member of the worker association.

Mafundi requires members to deposit KES 200 (USD 2) whenever they get a job. This amount is saved by the association and may be borrowed by a member when in social need. Some associations share the contributions weekly or monthly and also support members when welfare issues like sickness and death arise.

Training through apprenticeship is an important promotive service. Some individuals from formal construction firms and training institutions get into the industry as workers with skills and knowledge of the sector and become employers by sub-contracting labour, either as contractors or sub-contractors. Those from the unemployed pool get into the industry as novices, often taken care of by associations aligning the novices with experts. They begin as helpers taking any instructions, learning on the job, and if successful, advance into being semi-skilled workers with the potential of working alone and supporting other novices.

The dominant areas of training include building and construction, masonry, plumbing, electrical installation, and carpentry. Discussions with key informants revealed that electrical installation and, to some extent, plumbing are done in formal training institutions, while the other trades are largely learnt on the job and through apprenticeship. In FGDs, in Kisumu and Nairobi, it was noted that some contractors and sub-contractors provide mentoring job training programmes for different skill sets, as elaborated by a contractor, "I have done this for a long time since I started as a contractor ... some even coming from polytechnics. Some are contractors, others are *fundis*". Further probe revealed that such contractors work very closely with polytechnics and often get students attached to their firms on a regular basis. Several of such students continue to work with contractors after attachment, moving from one contractor or sub-contractor to another until they become independent construction workers, managers, and firm owners.

Associations rely on collaboration with other groups such as KCB, NYS, NITA, for free training, and legal advice. Some of the programmes done in collaboration with partners are advertised, while others go through associations and contractors to get trainees. The programmes are free and largely flexible which allow construction workers to continue working. Labour Link appreciated how the network had enabled the group to acquire new skills, as noted by a Key Informant and member of Labour Link in Kisumu, "...we have added to our skill set. We can do paper walling in a professional manner".

Transformative

Informal construction workers' associations in their ideal nature should provide agency for the workers through voice and representation. However, the findings reveal that transformative issues of rights, including negotiation for benefits such as health, pension, higher and stable income, protection against harassment, and negotiation of contracts, are not given high priority. The unpredictable work environment makes them largely concentrate on daily income, with low premium on issues of voice and representation that have potential of advancing their interest and transforming their work environment.

Having voice and effective representation requires active members of associations who participate and raise issues of concern such as lack of contracts and poor work environments. These are aspects which are generally not being actively negotiated, as noted by one of the sub-contractors in Nairobi "if you fail to complete the job because your client has not paid you, they will hire others to complete your job". This is partly due to lack of written contracts and lack of registration of workers and firms, which is exploited by contractors and clients who lack integrity. Unfortunately, having formal contracts does not prevent exploitation of workers. For fear of losing contracts, several workers and sub-contractors are not using the legal advice offered by KCB, to quote a sub-contractor in Kisumu, "when we get jobs we do not go for legal advice since clients do not like the legal way". One site manager pointed out that even the Chinese who have big projects do not give contracts. They rely on verbal contracts; "they do not hire labour but sub-contract to get labour" (KII: Site Manager, Kisumu). In this process, the written contract, if any, remains between the individual and the main contractor, while the workers rely on verbal agreement. This challenge is also linked to the Ministry of Labour's inability to enforce labour laws, especially in formal construction sites, and difficult judicial process.

In isolated cases, being in associations has encouraged a few workers to report to the leadership issues relating to non-payment and respect for agreements. Associations such as Migosi Builders undertake this task by negotiating with clients, and if not successful, the cases are referred to government authorities. The chairman of the association noted that most cases are resolved through negotiation by association, to quote, "it is good, because majority of cases are resolved ... It is because of being in a group". This infers that, although the findings show minimal use of transformative power, opportunity exists for use of voice and representation to improve the working environment of construction workers. Many workers do not bother to use such opportunities, and there is need for training and advocacy on the use of associations for leveraging transformative power.

Discussion

The associational life of construction workers is mainly centred on assisting workers to get employment opportunities and cushioning members by providing informal preventive and promotive social protection measures. Transformative social protection which entails engaging structures and institutions of power for changing their situation is marginally applied at case-by-case level; and not at a broader advocacy level directed at general conditions or the lack of enforcement of government rules. This is largely attributed to their weak organizational capacity which limits their ability to pool resources together to address structural, institutional, and societal issues facing the sector. This section discusses access to associations, highlighting entry requirements, comparing association membership status, and assessing access to formal social protection.

Access to associations

Workers' associations are powerful in assisting members to find job opportunities and supporting welfare needs, which makes them attractive to construction workers. This notwithstanding, only 33 percent are members of associations (Table 9.1).

Entry requirements

A large percentage of the associations restrict membership to individuals in the construction sector. Type of work is the dominant entry requirement followed by minor share for geographical location.

Discussion in FGDs and with KIIs in Nairobi and Kisumu confirmed the survey findings that membership is often based on working in the building and construction industry. A few associations like Migosi Builders admit those who do not work in the sector and also have no qualification for the industry but are keen on being nurtured into the sector. The associations also require potential members to be introduced by a member, in addition to availing Identification Card (ID), providing an address and name of their local administrator or chief. These requirements are aimed at providing points of reference for associations and ease traceability in cases of disappearance of workers and ensuring integrity and accountability of members. This is because the work environment is full of client assets which have to be protected and accounted for.

A mandatory membership payment is a requirement which cuts across all associations. In some associations, a requirement for a specified share contribution is embedded in the membership fee, whereas, in high level associations, such as the FundiTech Services Cooperative Society, there are higher requirements. Potential members have to be technicians, and certificate holders trained in specific skills and specializations in the building and construction industry. The cooperative requires potential members not to belong to any other cooperative society carrying out the same activity, members must buy some share capital, and have a three-month probation period. The latter is used to assess a member's character and performance.

Although most associations require members to be 18 years and above, one group, Dunga Builders association, requires members to be mature, with emphasis on marital status as a show of maturity – this was partly problematic to single mothers. A probe on this requirement revealed that it was aimed at protecting the group from those who exit after having a lump-sum payment before a full circle of savings benefits go to all members. It seems the requirements for access are generally simple, except for the service cooperative association which is targeting professionals in the sector. Other associations welcome both qualified and non-qualified workers. The other attributes of ethnic and religious backgrounds and being known to group members are not major requirements, but largely provide connection to associations through introduction.

Comparison of association membership status

Associations provide access to collective goods and are known to be of great importance to informal workers. McCormick, Mitullah, and Kinyanjui (2003), in a study of micro and small-scale enterprises (SMEs) in Kenya, observed that SMEs can collaborate and achieve what they cannot do alone (see also Haan, 1995). Members of associations are significantly more likely to be enrolled in formal SI schemes (Table 9.3) with over one third (36 percent) of association members having SI as compared to 25 percent of non-members.

Table 9.3 Differences in key workers characteristics by association member status

	Member	*Not member*	*Difference*	*t-Value*
Formal social insurance enrolment	0.36	0.25	0.11	1.78*
Health insurance coverage	0.39	0.32	0.08	1.19
Nairobi	0.56	0.55	0.01	0.12
Kisumu	0.44	0.45	-0.01	-0.12
Gender (male = 1)	0.93	0.88	0.06	1.45
Married	0.87	0.71	0.16	2.92***
Local	0.31	0.14	0.17	3.06***
Age	35.79	35.75	0.04	0.03
Mean daily earnings (current USD)	11.58	10.10	1.48	1.26
Assets (house and/or land)	0.28	0.31	-0.03	-0.43
Primary incomplete	0.06	0.09	-0.03	-0.90
Primary complete	0.43	0.35	0.08	1.20
Secondary or above	0.51	0.56	-0.05	-0.70
Training course	0.39	0.31	0.09	1.33
Training on job	0.50	0.46	0.03	0.45
Self-taught	0.11	0.23	-0.12	-2.30**
Wage-worker	0.97	0.92	0.05	1.67*
Own-account	0.02	0.04	-0.02	-0.75
Micro-business	0.01	0.04	-0.03	-1.58
Observations	221			

Source: Author's elaboration based on the survey data.
Note: <0.01***, <0.05**, <0.1*.

Association members are also more likely to be married and locally born. None of the education and training variables differ by association membership status, apart from self-taught workers being more likely not to be in an association. Non-members usually begin as helpers working with those registered with associations, learn the trade, join associations, and scale up to take professional positions in the industry. By contrast, wage workers are more likely to be in associations and are comparatively established.

Access to formal social protection

As also shown in Table 9.4, association membership plays a role in access to formal SI schemes since, out of those that have SI, 58 percent are association members, compared to only 45 percent of those that do not have SI. In addition, the table reveals that those workers more likely to be enrolled in social protection are male, married, have access to loans, have higher mean income and assets, and are more likely to have completed secondary school. Being an own-account worker

Table 9.4 Differences in key workers characteristics by social insurance enrolment

	Social insurance	*No social insurance*	*Difference*	*t-Value*
Association member	0.58	0.45	0.13	1.78*
Sacco/vicoba/chama	0.24	0.17	0.07	1.17
Work related association	0.29	0.25	0.04	0.66
Women/youth/religious	0.05	0.03	0.02	0.76
Work-related	0.24	0.25	-0.01	-0.04
Loans	0.29	0.15	0.14	-2.44**
Voice and representation	0.02	0.00	0.02	-1.54
Social cushioning	0.05	0.05	-0.01	-0.19
Nairobi	0.62	0.53	0.09	1.26
Kisumu	0.38	0.47	-0.09	-1.26
Male	0.97	0.88	0.09	2.15**
Married	0.86	0.75	0.11	1.81*
Local	0.18	0.24	-0.06	-0.93
Age	36.55	35.45	1.10	0.79
Mean daily earnings (current USD)	12.60	10.05	2.54	2.00**
Assets (house and/or land)	0.39	0.25	0.14	2.14**
Primary incomplete	0.05	0.08	-0.04	-1.01
Primary complete	0.30	0.43	-0.12	-1.72*
Secondary or above	0.65	0.49	0.16	2.21**
Training course	0.50	0.28	0.22	3.14
Training on job	0.36	0.53	-0.17	-2.27**
Self-taught	0.14	0.19	-0.05	-0.91
Wage-worker	0.92	0.95	-0.03	-0.92
Own-account	0.05	0.02	0.03	1.09
Micro-business	0.03	0.03	0.00	0.19
Observations	221			

Source: Author's elaboration based on the survey data.
Note: <0.01***, <0.05**, <0.1*.

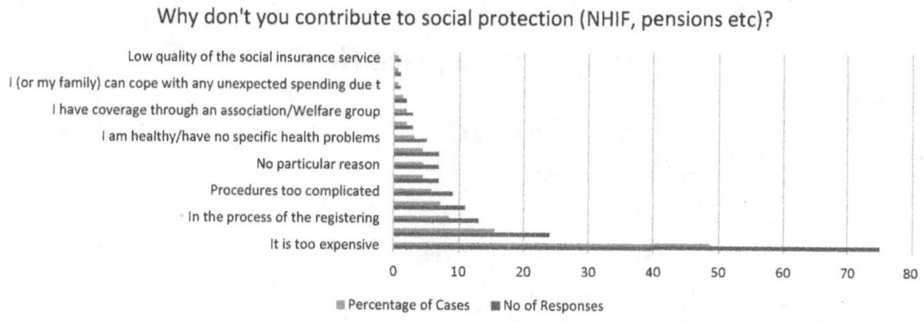

Figure 9.2 Reasons for not contributing to SI.

or working in cities where one is born does not make a significant difference in terms of social protection enrolment.

Higher mean income indicates that members of associations are comparatively better able to pay for and access formal social protection schemes. Several associations were noted to encourage members to register for NHIF, with some arguing that it relieves the association of the burden of taking care of members when they are sick.

Most of those registered with formal schemes were married with some being registered through a spouse. The survey findings indicate that cost, time, and inability to give the scheme attention are the main obstacles to enrolment with formal SI programmes (Figure 9.2).

A probe through FGDs and KIIs in Kisumu and Nairobi for reasons for not subscribing to formal SI schemes confirmed the survey findings. Most workers consider the KES 500 payments per month expensive, especially when they consider that, in many cases, they are healthy and do not use the service. The bureaucratic processes including restriction of outpatients to one hospital are further drawbacks to construction workers who are constantly moving in search of employment. Thus, the study revealed a remarkable deficit in formal SI among informal construction workers.

A further probe showed that 14 percent of workers who previously subscribed to NHIF stopped due to mixed reasons dominated by too expensive, followed by leaving previous job, school, college or divorce, not having made use of the scheme, lack of stable job and remittances, and procedures too complicated (Figure 9.3).

The SI deficit in the sector is explained by informality and the fragile nature of work, in which there is no specific person or employer to take responsibility for workers. Furthermore, both workers and contractors do not take insurance seriously, as highlighted by a KII Contractor (Box 9.3) and other voices of construction workers drawn from FGDs in Nairobi and Kisumu.

The quote from the constructor reveals that accidents are treated as bad luck. A probe through FGDs and KIIs in Nairobi and Kisumu on the issue indicated that liability should be shared between the contractor and the client, but often, the two do not take full responsibility for workers. In most cases, they take the injured to hospital, make a one-off payment, and leave the employee in the hands

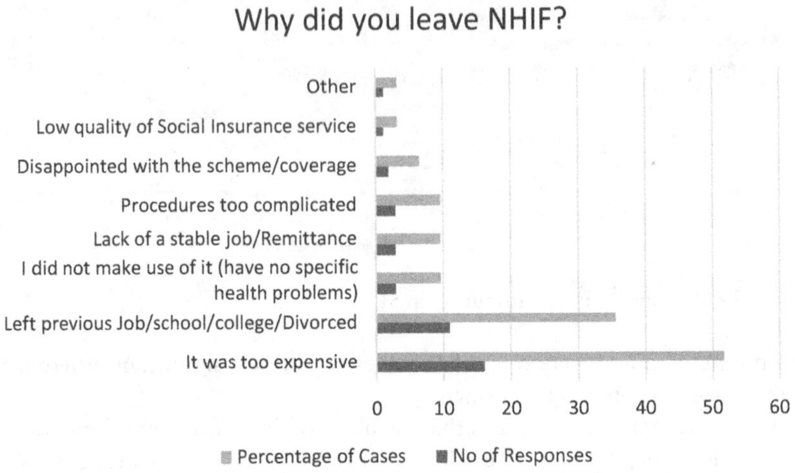

Figure 9.3 Reason for leaving NHIF.

Box 9.3 Contractor's view and other voices of workers on social protection

Contractors take insurance for granted except in isolated cases where they work for government or big jobs, but for small contract that last for two weeks, three weeks, and you will not be with them for long, we do not insure them ... but we pray to God so that a bad incident does not occur ... that is what we pray for ... if accidents happen then God may have planned that you die like that or get hurt like that. Most contracts have been completed without a hitch, including the ones in which workers were insured. There is a job we did in Milimani area in Kisumu, and we secured insurance for all workers and concluded it well without any incidents (KII, Contractor, Kisumu).

I had initially enrolled for NHIF and paid for three years without falling sick. I did not see the need as I felt that the money was not helpful. After some time, I got married and I have not had any case of sickness so I stopped contributing. I joined M-TIBA which has a minimum of KES 100 (USD 10) monthly, and I find it a better option (FGD Participant, Nairobi).

We formed a group to take care of health insurance especially sickness. I have seen my friends being fined for default and I have lost hope in the cover. I now go to mission hospital. As a group, we have emergency fee for accidents occurring on-site. Our Constitution clearly states that the nuclear family will be covered in case of sickness and death, we are building our own insurance ... we assist members until they get well (FGD Participant, Nairobi).

> ... I can go for a week without a job. It becomes hard for me to accumulate KES 500 (USD 5) at the end of the month. It is easier for the formal workers because theirs is deducted from monthly salary (FGD, Nairobi).

Notes: The M-tiba is an e-Wallet. If one seeks health services, they are covered only for the contributed amount, and any balance would be paid out of pocket.

of relatives, colleagues, and friends. Some of the contractors and clients also assume that the responsibility of having the right gears and being safe totally lie on the employee, which goes against the WIBA requirements.

The deficit in formal SI schemes is filled by informal insurance provided through associations, relatives, colleagues, and friends, which are bottom-up loose strategies suitable to the workers. Almost all associations had an SI welfare component, ranging from an informally defined contribution to ad-hoc contributions when needs arise. In the former, members contribute a specific amount of money per month with a percentage dedicated for SI. They also run savings and credit schemes which allow members to get back some portion of their shares through revolving saving and credit schemes enabling easy borrowing. This aspect and informal models of SI are very attractive to construction workers.

Both the defined and ad-hoc contributions allow members to save and accumulate money for SI, largely assigned to sickness and death. These models are preferred compared to subscription to formal SI schemes, which are viewed as complex and bureaucratic. Some of the workers were paying more money in their associations, compared to what they would pay for formal SI. The association payments take part of the income which would otherwise pay for formal social protection because the workers trust associations compared to public schemes. There is therefore a need to address the work environment, bureaucratic, and technical procedures that keep informal construction workers from subscribing to formal public SI schemes.

Conclusion

Informal construction workers are weak in associational life and have not managed to fully leverage the various powers latent in existing associations. Most of their operations are horizontal, engaging with membership, and other construction workers outside their associations largely in search of employment opportunities. Although they are interacting with public and private agencies who provide training, this interaction is limited to one-off training without exploiting other potential powers which such interaction can leverage. This is largely due to weak organizational capacity of associations and the construction workers' work environment which does not leave adequate time for engaging in associational life and tapping other powers for formal social protection.

Most associations in collaboration with family members, relatives, colleagues, and friends take care of construction workers' social protection needs such as sickness, death, and other social livelihood needs. Cases of OHS are handled in a haphazard manner with many workers left on their own once admitted to hospital or given the first treatment after injury on-site. Many workers do not perceive SI as a right in spite of existing institutional measures that have been put in place by the state. Their informality is also exploited by employers, including formal firms that should hire workers under contracts and ensure decent work. This problem is further intensified by poor state governance and enforcement, which include a lack of resources that prevents government labour office inspectors from undertaking inspections on construction sites.

The associations which are supposed to provide agency for claiming SI rights and protecting workers are equally weak and require capacity-building and exposure on how SI, including NHIF and OHS operate. Basic facts on how insurance operates are not well known to the workers and partly contribute to their thinking that one must use the service for the scheme to be meaningful.

Addressing these deficits requires going beyond associational power to using structural, institutional, and societal powers to negotiate the work terrain and related issues. Associations largely concentrate on preventive and promotive social protection measures, focusing largely on welfare issues, savings and credit, and providing information on job opportunities but hardly address work environments in the sector. Cases of accidents on-site are handled by foremen, colleagues, and relatives as opposed to using legal workplace protective measures. The deficits in SI measures are not viewed as the responsibility of clients and contractors, but that of workers.

The work environment, associational weaknesses, and technical bureaucratic issues undermine formal SI programmes targeting construction workers. The majority of workers do not access the schemes due to cost and poor design, including requirement for payment within a specific time, default attracting penalty fees – which workers find punitive – and inability to access contributions in form of savings. These challenges push workers to opt for associational life and its informal models of SI which provide inadequate cover, albeit easily accessible. The associations might be persuaded to divert some of the savings they share in their groups to formal SI which has more comprehensive cover compared to informal models. This should be complemented with other products designed for informal workers such as M-TIBA and Mbao pension schemes, which seem to be more acceptable to the informal workers, and also provide an opportunity to rethink how the informal workers can be brought under the umbrella of public SI schemes.

Overall, most challenges faced by construction workers are shrouded in informality, and yet, the literature points out that the formal and informal sector have synergy and should be addressed as a continuum as opposed to a silo approach that dismisses one sector and upholds the other. Application of a two-model social protection approach which uses both bottom-up efforts of construction workers and top-down public formal models is likely to advance the situation of informal construction workers. This should entail taking into consideration the nature of construction industry including its unique irregular operations, multi-layered nature, and the lack of legal contracts. The latter seems to be a major problem, exploited by clients and contractors, a challenge that can be addressed through strong associations with the ability to leverage all sources of power for the development of effective social protection measures and overall development of the sector.

Notes

1 Unless otherwise noted, the average USD rate from June 2018 to December 2018 (USD 1 for KES 101.2) is used throughout the chapter.
2 Loose networks and associations are not formally registered, members do not pay registration fees, and meetings are ad hoc and largely driven by availability of work or connections for work.

References

COSS (2020) *Code of Social Security, No. 36 of 2020.* Ministry of Law and Justice, Government of India. Delhi, India.

Devereux, S. & Sabates-Wheeler, R. (2004) *Transformative social protection.* IDS Working Paper 232. Brighton, Institute of Development Studies.

Haan, H.C. (1995) *Micro and small enterprises associations in Africa.* TOOL Consult Working Paper 1. Amsterdam, TOOL Consult.

McCormick, D., Mitullah, W.V. & Kinyanjui, M. (2003) *How to collaborate: Associations and other community based organisations among Kenyan micro and small-scale enterprises.* Institute for Development Studies. Nairobi, University of Nairobi.

Mitullah, W.V. & Wachira, I.N. (2003) *Informal labour in the construction industry in Kenya: A case study of Nairobi.* Geneva, International Labour Office.

ROK (2007) *Vision 2030: A globally competitive prosperous Kenya.* Republic of Kenya. Nairobi, Government Printer.

ROK (2010) *Constitution of Kenya.* Republic of Kenya. Nairobi, Government Printer.

Schmalz, S., Ludwig, C. & Webster, E. (2018) The power resources approach: Developments and challenges. *Global Labour Journal.* 9 (2), 113–134.

Shepherd, A., Marcus, R. & Barrientos, A. (2004) *Policy paper on social protection.* Overseas Development Institute Report. UK, Department for International Development (DfiD).

Silver, B.J. (2003) *Forces of labor: Workers' movements and globalization since 1870.* Cambridge, Cambridge University Press.

Wells, J. (2007) Informality in the construction sector in developing countries. *Construction Management and Economics.* 25 (1), 87–93.

Wright, E.O. (2000) Working-class power, capitalist-class interests, and class compromise. *American Journal of Sociology.* 105 (4), 957–1002.

10 Convergence and divergence of workers' environment, associations, and access to social protection

Sectoral and country comparisons

Winnie Mitullah, Lone Riisgaard, Nina Torm, Aloyce Gervas, Raphael Indimuli, Anne W. Kamau and Godbertha Kinyondo

Introduction

The literature on the informal economy tends to make generalizations across sectors and countries. While this is true to some extent, the comparison between Kenya and Tanzania reveals both convergence and divergence depending on the issue of investigation. The two countries have different historical development trajectories, which are also reflected in the development of the different industrial sectors and in associational life. On the other hand, Kenya embraced the spirit of *harambee* (pulling together) right at independence, an approach which enabled many workers and families to cushion themselves in a capitalist liberal economy which was less concerned about those who could not fend for themselves. On the other hand, Tanzania had the *Ujamaa* modelled on an African collective approach but steered by the state that ensured that basic services were available to citizens, albeit in a limited manner.

These differences have had an effect on growth and development of the two countries, including the development of the informal sector. The capitalist state in Kenya nurtured a society of gross inequalities with a large informal sector, while the Tanzanian state had comparatively fewer inequalities and a comparatively smaller informal sector. Consequently, Kenya has had a longer experience handling informality, dating back to the famous 1972 ILO report (ILO, 1972) focusing on employment, incomes, and inequalities in Kenya. This long experience is demonstrated by transport SACCOs which are operating as quasi-legal entities working closely with the government in governance of the sector, while in the Tanzanian transport sector, cooperation is informal. The trade sector in Kenya is following a similar pattern to the transport sector with representation of informal traders in the Micro and Small Enterprises Authority (MSEA). Tanzania does not have such formal institutional linkages between informal workers

DOI: 10.4324/9781003173694-10

and the government, although intermittent support for the informal sector and ad-hoc coordination and consultation does exist.

One aspect which cuts across the three sectors covered in this study and across countries is the work environment. Informal workers operate under difficult and precarious working conditions often without contracts, social protection, and labour rights. Of particular concern is the lack of contracts among workers who work in an identifiable employment relationship, in particular, the transport and construction workers. This undermines their chances of accessing labour rights as provided in both Kenya's and Tanzania's policies and regulations. Although the majority, especially the micro-traders, operate as own-account workers, even in cases where workers are employed by firms such as in the construction sector, the multi-layered nature of employment subjects workers to sub-contracting without individual contracts and security. The situation is not different in the transport sectors. In Kenya, the formation of transport SACCOs and franchises has not resolved the issue of contracts, and – where they have contracts – they are not enforced, resulting in an inability to claim and access labour rights (Table 10.1).

Another aspect which cuts across is a rich associational life, where workers' collective associations provide a range of services to their members, many of which can be considered informal social protection measures. As seen throughout this volume, the services provided by associations differ markedly across sectors (and by country), which testifies to the importance of sector-specific insights. Yet, common across sectors in both countries is that entry into associations requires payment of fees and adherence to group constitutions or regulations.

In order to compare and analyze sectors across countries, four major issues are discussed in this chapter: (i) How key worker characteristics relate to workers' access to formal and informal insurance by country and sector, (ii) How governance frameworks across countries and sectors function, including associational differences and power resources available to associations, (iii) Whether unions facilitate representation of informal workers, and (iv) How associations provide preventive, promotive, and transformative social protection measures to their members in line with the framework developed by Devereux and Sabates-Wheeler (2004). Each of the corresponding sub-sections teases out convergences across the sectors in Kenya and Tanzania, whilst bringing out unique aspects pertaining to each sector in each country. Last, the conclusion pulls out the key findings, noting the difference and similarities between the countries and the sectors.

Table 10.1 Formal social insurance coverage (percent)

	Construction	*Trade*	*Transport*	*Overall*
Kenya	35	39	50	41
Tanzania	17	22	16	19

Source: Author's elaboration.

Who are the informal workers?

On a global scale, the rate of informality is higher among men than among women; yet, in low- and lower-middle-income countries, women are more likely than men to be in informal employment (ILO, 2019). For instance, in Africa, 90 percent of employed women are in informal employment in contrast to 83 percent of men (ibid.). In the project on which this volume is based, a higher share of women informal workers is found among the micro-traders in both Kenya and Tanzania, whereas the construction and transport sectors are dominated by young male workers. In the Kenyan construction industry, however, the male–female ratio seems to be changing slowly, as women are increasingly taking on special skills such as painting, electrical works, and masonry. Likewise, in transport, women are also gradually entering the sector, working mainly as conductors, yet almost absent among the motorcyclists.

Across the three sectors, we find that several worker characteristics differ along the gender dimension, although with some variation between Kenya and Tanzania. For instance, in Tanzania, women are more likely to contribute to formal social insurance, whilst in Kenya, they are less likely, and when it comes to association membership, this is more common among women in both countries. Women have a history of having barriers to formal credit facilities, and associations have always been their source of support and access to credit. In line with the general wage gap literature, male workers have higher average earnings, although the difference is more pronounced in Tanzania where women are more likely to possess assets. Across both countries, men are more highly educated and also more likely to attend training courses, whilst women are generally self-taught. Regarding worker types, in both countries, men dominate amongst wage-workers whilst women tend to be own-account, and in Tanzania, women are also more likely to be micro-firm owners. In Kenya, there is a variation in the transport sector where both men and women are wage-workers. Own-account operators are few, and when women join the sector, they work mainly as conductors while men occupy the better-paying jobs of drivers. These gender-divisions point to women having to combine reproductive work with their income-generating activities, and in the case of the transport sector in Kenya, it is also due to late entry of women in the sector, mainly considered to be men's work and a risky occupation. Interestingly, when looking at the challenges mentioned by informal workers, there are no major differences by gender; yet, Kenyan women workers mention working conditions and safety issues as more pressing, whilst Tanzanian women cite payment delays and few clients and/or competition as main concerns. For men, issues related to harassment/arrest/police cut across the countries, as do payment delays, and common to all workers across the countries are credit concerns. Finally, in terms of the benefits of association membership in both countries, women and men equally cite financial and social support as the most important which is in line with the credit challenge they face.

Apart from the gender dimension, workers in the informal economy differ widely along a number of other characteristics including education level, income, associational belonging, and social protection, and such differences are also reflected in our survey data. In terms of education, across the sectors, Tanzania has a much lower share of workers with completed secondary education (29 percent) compared with the share of workers that have primary education (62 percent). By contrast, in Kenya, the incidence of completed secondary education (and above) is at 47 percent, slightly higher than the share of workers with completed primary education (42 percent). As expected, our figures are lower than official secondary education enrolment figures for the entire population, which are 71 percent in Kenya (ROK, 2019a) and 32 percent in Tanzania (URT, 2019, p. 255). Finally, as also seen in Chapter 3, participation in formal social insurance is much more likely for workers with secondary education and above, whilst education level is not a determinant of association membership.

Regarding income, we also find variation between the two countries and across sectors. As expected, given the general difference in wages and living standards between the two countries, median daily earnings are higher in Kenya compared with Tanzania among both traders and transport workers (see sector-specific chapters for further detail). However, when it comes to construction, Tanzanian construction workers have median daily earnings that are almost on par with Kenyan workers, which, to a large extent, can be explained by the types of workers interviewed. For instance, in Tanzania, the construction workers included a higher share of masons, whilst, in Kenya, the majority sampled were helpers who are lower on the earnings hierarchy compared with masons who are the ones negotiating for jobs with sub-contractors, contractors, and clients and hiring helpers to assist. Relatedly, the informal worker earnings observed in Tanzania are comparable to those of formal sector workers, whilst, in Kenya, official monthly earnings are substantially above the informal earnings reported. When it comes to participation in formal social insurance, unlike education, this is not determined by earnings. However, in Kenya, association members have higher average earnings than non-members, whereas, in Tanzania, the opposite is the case (see Chapter 3 for more detail).

Finally, in terms of key informal worker challenges and benefits of associations, there are no major differences by income level, except that, in Tanzania, those mentioning the opportunity to save as a top benefit also have lower average earnings, which seem logical. In accordance with the general literature, we find that earnings increase with age (up until a certain point) as it is a measure of experience/seniority. Moreover, across sectors and all things being equal, worker age is correlated with an increased likelihood of listing health benefits (over voice and representation, work-related issues, and loans), whereas more educated workers list voice and representation over health as a main association benefit.

In terms of geography, construction workers in Tanzania have a higher percentage of workers operating in cities where they were born compared to Kenyan workers who were largely moving in search for jobs. This is partly explained by the higher (37 percent) level of urbanization in Tanzania (UN, 2019) compared to Kenya (28 percent) (ROK, 2019b). Furthermore, since most Tanzanian workers work in towns where they were born, they have higher chances of owning property either by inheritance or buying land which is cheaper in Tanzania compared to Kenya. This difference is reflected in the data where 43 percent of Tanzanian workers own assets compared with 27 percent of Kenyans. Overall, association members have more assets than non-members, which could be either a reflection of more wealthy workers selecting to be in associations or the latter facilitating and/or providing financial support to their members.

In addition to the trends in gender, education, income, and worker-type characteristics, workers in the informal economy differ substantially in terms of their associational belonging, as seen throughout this volume, and discussed further in this chapter. However, on this dimension, one aspect that cuts across countries and sectors is the important role played by different workers' associations in filling the substantial formal social insurance gap by assisting members in accessing employment opportunities and cushioning members during difficult times such as sickness, death, and social welfare needs.

Institutional and governance framework

The deficits in contracting and social protection among informal workers prevail despite the varied sectoral institutional and governance frameworks that exist in each of the countries, albeit with marginal differences. Kenya is comparatively more advanced in putting in place institutions to govern the transport and trade sectors, while both countries are on par in respect to the construction sector. While having formal state institutions is expected to improve the general work environment, some of the institutions are largely not in harmony with the informal working environment and institutions of workers such as associations and networks which drive workers' operations on a daily basis. This is glaring in the area of social protection where many workers do not enrol in formal social insurance (SI) schemes due to complex bureaucratic procedures for accessing services and irregular incomes. Instead, they largely rely on associations for social protection due to the close reach and flexible non-bureaucratic processes. Thus, this section examines the formal governance institutions for the three sub-sectors across countries, and power resources and related applicability to formal workers.

Formal institutions

An examination of the three sectors reveals deficits in application of public policies, regulations, and standards. All informal workers can, privately or through registered associations, enrol voluntarily in the National Hospital Insurance Fund (NHIF) in both Kenya and Tanzania, and in Tanzania, the same arrangement is also available through the community health fund (CHF) under the NHIF. The KIKOA scheme which previously provided for group registration of informal workers through their associations in Tanzania had potential as it attracted considerable interest from informal workers' associations, but it did not survive partly due to adverse selection problems.

Although recent years have seen an expansion in informal membership rates in public schemes (see Chapter 2), overall, the latter seem not to work well for informal workers, as shown in the relatively low enrolment of workers participating in formal social insurance (health or pension). For the vast majority of workers in both countries, formal social insurance refers specifically to health insurance, and in both countries, individual NHIF enrolment also covers immediate family members. Thus, the measure of health insurance coverage provides a more accurate depiction of actual coverage, and according to our data, this is 41 percent for Kenya compared with 19 percent for Tanzania. These figures are in line with the official coverage rates which are around 39 percent for Kenya (KNBS, 2019) and 22 percent for Tanzania (Jacob & Pedersen, 2018). The general picture of a much higher formal social insurance coverage among informal sector workers in Kenya is reflected at the sectoral level; yet, the difference is particularly stark in the transport sector, where 50 percent of Kenyan workers have health insurance coverage compared to only 16 percent in Tanzania, as seen in Table 10.1.

This difference in the transport sector is due to the specific set-up for informal transport workers in Kenya, whereby some bus companies/franchises or SACCOS commission certain savings to pay for the workers' NHIF (and NSSF) coverage. However, even in this case, membership remains largely voluntary. Additionally, for Kenya, the introduction of UHC in pilot counties – including Kisumu, one of the volume case studies – and intensified recruitment-drive by NHIF, could account for the higher Kenya figures. Further examination of the three sub-sectors shows convergence and divergence, as examined below.

In the **construction sector**, legislation such as Work Injury Benefit Act (WIBA), Occupational Safety and Health Act (OSHA), and Occupational Health and Safety (OHS) rules in Kenya and Workers Compensation Fund (WCF) in Tanzania are not being complied with. These institutions largely work for those formally employed, and yet, the majority of construction workers are operating informally. This has exposed the workers in both countries to exploitation by the industry, with formal enterprises employing workers on casual basis to avoid costs associated with employment. This leaves the workers with informal social protection schemes based on associations, relatives, friends, and good Samaritans as the only fall-back.

In the micro-trade sector, both countries, during the last few years, have seen less confrontations and conflicts between micro-traders and authorities, particularly in Kenya. The 2010 Kenya Constitution (ROK, 2010) takes a rights perspective, highlighting rights to social security, protection for marginalized groups, and participatory governance. One outcome of this has been the Micro and Small Enterprises (MSE) Act of 2012 which provides representation of micro-enterprises including micro-traders (see Chapter 7). In Tanzania, the change seems to be mainly due to the more positive attitude of the late president Pombe Magufuli (see Chapter 6). In comparison, the changes in Kenya are institutionalized in the legislation, whereas in Tanzania it is not and hence rather dependent on politics of the day and the approach of a particular politician at a given time.

The transport sector, with its many actors and institutions, poses a challenge in respect to organization for social protection. The presence of a large informal workforce makes it difficult for the Tanzanian and Kenyan governments to regulate the sector, even though there are efforts – mainly in Kenya – to organize the sector through SACCOs and franchises. For instance, since 2010, it is a legal requirement for Kenyan matatu (PSV) workers to belong to a SACCO or a transport management company. The workers, however, remain informal and have daily remittance targets (Spooner & Manga, 2019) that force them to work for long hours, to aggressively compete for passengers, and drive recklessly in order to increase the number of trips made per day, hence compromising on safety. In Tanzania, local authorities encourage the formation of associations in the transport sector for easier coordination, and rule-bound cooperation does exist, even if it is not institutionalized.

In sum, across sectors, the formal social insurance coverage gap in both countries is partly due to a lack of understanding of the importance of social insurance, coupled with meagre irregular earnings, the bureaucratic processes involved, and the structure of the schemes. In Kenya, this is also attributed to the government's failure to enforce regulations requiring transport workers to have social protection. Furthermore, formal social insurance schemes are mostly and historically designated to deal with the formal sector, thus lack capacity to deal with a more challenging but expanding informal sector. In the following sub-section, we discuss the power resources used by associations, and in the section thereafter, we provide examples of different types of informal social protection measures/services that associations offer their members.

Associational power resources and the applicability to informal workers

Informal workers and their associations in both countries wield associational, institutional, structural, and societal powers which they use to negotiate their work environment and livelihoods. As discussed in more detail in Chapter 1 of this edited volume, the PRA starts from the basic premise that, if organized, labour can successfully defend its interests by collectively mobilizing different power resources (Schmalz, Ludwig & Webster, 2018). The approach was however developed to analyze the power resources available to trade unions – not informal workers. This begs the question of how useful the power resource approach is to cast light on informal worker associations. We argue that, if modified and employed with caution, the analytical typology still has explanatory power, particularly in highlighting how sector- and country-specific groups of informal workers have differing power resources available – specifically with regards to the transformative social protection measure of voice and representation – as discussed below. The applicability of the approach and the implications of applying it to informal workers will be discussed in more detail in the concluding chapter of this volume.

Associational power

Associational power is derived from the formation of collective organizations of workers and, as mentioned, can lead to transformative social protection measures in the form of voice and representation. Given the context of informal workers, we included all the different types of collective associations created or engaged in by informal workers to advance their own interests and conditions. It is an open empirical question who might form relevant counterparts with regards to representation, as the case studies revealed instances of representation towards employers (in transport and construction), the police (transport and trade), as well as municipal and national level authorities (all sectors). Associational power understood in this broad manner is potentially of immense importance for informal workers. Other than numbers, the strength of associational power is associated with a number of factors such as effective organizational structure, member willingness to take action such as participating in campaigns, willingness to pay dues, and a strong collective identity. It is hard though to generalize about informal worker associations since, as illustrated throughout this book, they differ significantly in particular in the following parameters: resources, representational power, and key purpose of associations. Moreover, as seen in Table 10.2, our survey data shows that the incidence of association membership varies substantially across the three sectors and between countries.

Overall, a higher percentage of informal workers are enrolled in associations in Kenya compared with Tanzania, and informal transport workers have a stronger associational life compared to informal traders and, in particular, compared to construction workers in both Kenya and in Tanzania. Hence, based on numbers, associational power differs substantially between sectors, and in general, Kenyan associations were often found to be more resourceful.

Nonetheless, there are more similarities in associations than differences when comparing the same sectors across countries. Workers' associations are structured to perform functions required by members, which are mainly assisting to find jobs and cushioning members during difficult times for construction workers in both countries; welfare and protecting trading sites for traders in both countries; government requirement for business, welfare, and negotiation for Kenya transport workers; and welfare and negotiation for transport workers in Tanzania. Interestingly, as the analysis in Chapter 3 shows, there is a significant correlation between informal association membership and access to formal

Table 10.2 Informal association membership shares (percent)

	Construction	*Trade*	*Transport*	*Overall*
Kenya	33	47	57	48
Tanzania	19	34	50	34

Source: Author's elaboration.

social insurance, which holds across sectors and when other factors like earnings, education level, and location are taken into account. This relationship is stronger in Kenya where the share of association members that are enrolled in formal schemes is double at 40 percent compared with 21 percent in Tanzania; yet, further analysis reveals that the correlation between informal association membership and access to formal social insurance is mostly seen among transport workers in both Kenya and Tanzania.

In both the survey data and interviews, large sectoral differences were revealed with regards to representation and the key purpose of the associations. In terms of key benefits of associations, at a general level, the survey data reveal that, controlling for key worker characteristics, trade and transport workers list voice and loans as more important (compared with health benefits) relative to construction workers (i.e. the latter are relatively more likely to list health benefits as key association functions).

Most associations encountered in the **construction sector** were created to facilitate job-seeking while also providing cushioning for members and only a few worker associations were engaged in bargaining with employers. More specifically, a few associations in Kenya were found to represent members to raise voice and demand their rights, including demands for work-space, fighting harassment by local authorities, and demanding payment for workers' dues from subcontractors, contractors, and clients. In the survey, issues like wage delays and short contracts were listed as relatively more important challenges for construction workers, all things being equal. The construction sectors in the two countries differ slightly. In Tanzania, where members are actively participating in association meetings and contributing, associations are stronger. Old associations tend to be dormant while younger ones are more active. This can be partly explained by the fact that old associations may have achieved their intended goals. Furthermore, Tanzania associations do not have umbrella associations, while Kenya has umbrella associations, but very few workers belong to such associations. No informal construction workers were found to belong to unions, although there were earlier attempts in Tanzania to organize workers to join trade unions through the support of development partners, but this was not sustained. This is despite the existence of such associations, as evidenced in the study.

In transport, in general, workers are mainly organized around work-related challenges, particularly problems in dealing with authorities or employers while some associations are also created for the purposes of social welfare and protection. Correspondingly, the survey data shows that, for transport workers, key challenges include dealing with the police/authorities, client conflicts, and poor infrastructure. Transport associations in both countries present a fragmented patchwork often including break-away associations counting amongst them a range of transport unions. In Kenya, as mentioned earlier, legislative provisions have made the associational landscape quite particular for the matatu sub-sector.

A key directive of the legally mandated SACCOs and transport companies is to negotiate with authorities on issues affecting the sector. These include fleet operations, management, and responding to government directives. However, the issues addressed are mainly related to factors which might impede a smooth operation of the fleet and are not as such related to working conditions or rights to social protection. Hence, these associations mostly favour the interests of SACCOs and vehicle owners while paying little attention to the issues affecting workers. Therefore, they cannot really be considered vehicles for workers' associational power which would allow them to voice their concerns or to demand for better working conditions. Nonetheless, along with some of the well-functioning associations, some SACCOs and companies invest in workers' capacity-building by organizing training on a range of issues that include road safety, financial literacy, and investing in social protection among others.

It is in the **micro-trade sector** that we probably find the largest variety of associations varying from groups working in designated areas who organize access to work-space and market-level representation, over welfare groups, to larger and more robust umbrella structures organizing many different kinds of traders. With regards to associational power, what stood out from the research, particularly on traders, was whether the association belonged to a larger umbrella structure (e.g. VIBINDO in Tanzania and KENASVIT in Kenya). In general, associations linked to umbrella associations were able to offer different kinds of support to their affiliated groups, such as access to training and infrastructure to take up issues with regards to representation. By contrast, smaller stand-alone associations would often be more inward-looking. The larger and more robust umbrella associations were generally also the ones which would engage in voice and representation. In Tanzania, this included a trade union which is discussed in more detail in the last part of this chapter.

Institutional power

Institutional power is derived from laws, regulations, procedures, and practices that regulate the relationship between worker associations and employers as well as authorities. In the context of informal workers, sources of institutional power in our cases stem mainly from either specific provisions in the legal system or affiliation to trade unions, and we will discuss these in turn.

As already mentioned in the introduction to this chapter, Kenya has taken steps (albeit small) towards a more inclusive and participatory institutional set-up with the new constitution and particularly with the Micro and Small Enterprises (MSE) Act of 2012 which provides for representation of informal traders (represented via KENASVIT) amongst other groups. This means that, at least for the micro-trade sector, some form of representation is institutionalized.

In Tanzania, the change seems to have gone in the opposite direction with moves towards centralization and shrinking of civil society space more generally. As described in Chapter 7, a seat was given to a representative of an informal association on the National Employment Advisory Committee in 1999, but there is little evidence of the committee currently being active on informal sector matters. Hence, even though some groups of traders were able to use the established name and personal contacts of the leader of the umbrella association VIBINDO to access official structures, this has been in an ad-hoc manner with no institutional guarantee for consultation or representation.

The case of micro-traders further illustrates how access to trading space – whether illegal, legal, or simply tolerated – is sometimes the context on which institutional power depends. Hence, a low-level form of institutional power is available to associations of members who trade in a local government-sanctioned place. Such groups often have a form of representation in a market council, and through this, they have some access to negotiate with local authorities on behalf of their members.

In the case of informal transport workers in Kenya, we see another form of institutional influence in that the legal framework in place for matatu transport is very influential in regulating the relationship between the transport workers and the owners of the vehicles although not always in ways intended. Hence, while the workers require SACCOs' clearance in order to get the mandatory road service license and badges to operate in the sector, the SACCOS mostly favour the interests of vehicle owners in terms of securing the smooth operation of their vehicles rather than guaranteeing that workers' rights are respected or their opinions voiced.

For informal construction workers, the relationship between informal worker associations and employers (which are sometimes public entities) is at times institutionalized in that some of the better organized and resourced associations are able to bid for tenders and win formal contracts. This was visible in the case of FundiTech service cooperative in Kenya and Sinza Kijiweni Construction in Tanzania which were collaborating with local authorities for contracts. Labour Link in Kenya, a loose network of construction workers, had also nurtured relations with private firms and was able to get contracts.

In the transport sector in both countries and in the micro-trade sector in Tanzania, we saw informal workers tapping into established tripartite structures via affiliation with newly formed or established trade unions. In Kenya, the established transport sector union (TAWU) began to recruit amongst informal workers only in 2018, but, according to the secretary general, they had recruited around 1,000 matatu workers from five SACCOs and negotiated a collective bargaining agreement with one of them. They have been in the process of changing the constitution to better accommodate informal workers; yet, it remains unclear what rights were to be awarded to informal members internally in the union (Riisgaard, 2021). In addition, Kenya Matatu Workers Union (MWU), registered in 2013, organizes around 1,000 informal matatu workers. The key service they provide to members is to represent them when problems occur – for example, delayed payment from the employer or issues involving the police, but unlike the more established unions, MWU still does not have the capacity to represent members in the court system (ibid.). Further, the Public Transport Operators Union (PUTON) in Kenya represents workers in transport-oriented businesses, including the digital platform workers. In general, however, informal transport workers were found either not interested in joining the available unions or fearing victimization by employers, including SACCOs in Kenya.

In Tanzania, three different unions are of relevance to daladala and bodaboda riders. COTWU-T is the established trade union which – similarly to Kenya – has been weakened by breakaway unions. At the time of the interviews, COTWU-T had active engagement with three informal worker associations of which one consisted of clerks at a bus terminal while another one consisted of bodaboda riders. The last was a group of women fast food providers working at a bus terminal.

The informal associations are incorporated as branches, and the paying informal members are awarded the same rights within the union as formal members. COTWU-T also used to organize a larger group of daladala workers, but they have since broken out and registered their own union TAROTWU. Another new union is Tanzania Association of Drivers Union (TADU) that represents bodaboda workers and those that operate smaller vehicles like bajaji. The latter two unions, in cooperation with a union for long distance bus drivers, are attempting to form a federation of informal transportation workers' unions with the aim of representing informal transport workers as a distinct voice in tripartite forums. However, that kind of representation is currently limited to affiliates to the established federation TUCTA.

In Tanzania, some informal associations – primarily working in markets – have during the last decade or so been incorporated into the established trade union TUICO as affiliate unions. A newer initiative from TUICO, originating in an externally funded programme with StreetNet International, has established a bargaining committee with two members from each of the market-based branches with the aim of providing a bridge between the municipal authorities and traders. The increasing (although still very limited) incorporation of informal trader associations into trade unions has to some degree opened up an institutionalized space for representation and voice for informal traders, albeit as part of the trade union movement in the established institutional model and in addition through this innovative bargaining committee, although further research is needed to assess its current functionality.

In summary, very few informal workers are yet organized into trade unions. Most successful so far has been the transport sector where both new and established unions are involved, however in a very fragmented and competitive manner. In general, the recent opening towards informal workers poses a very difficult and politicized challenge for the established unions, as discussed in more detail by Riisgaard (2021). Notwithstanding the difficulties, inclusion provides an interesting potential for increasing institutional power by tapping into established institutional setups. In the conclusion, we discuss this in relation to different representational models and in comparison with the informal workers' grassroots associations.

Structural power

In line with Silver (2003), structural power can be divided into workplace bargaining power (the power that can be utilized from the strategic location of a certain group of workers in a key sector) and marketplace power (relating to the tightness of the labour market). Activities in the informal economy are most often characterized by low entry barriers, and since the availability of labour is vast, informal workers will, in general, command weak marketplace power. Nonetheless, *strategic disruption abilities* exist, although they differ greatly by sector. This is of key importance for informal transport workers who, if coordinated, can bring the city circulation to a chaotic halt. Even though transport worker associations are fragmented in both countries, their existence and the fact that they could create considerable disturbance if provoked is acknowledged by public officials.

The potential for workers to disrupt transport operations and hence the circulation of people and goods in the cities is real and has been witnessed in both countries, for instance, to demand for improvement of road infrastructure or protesting police harassment and crackdowns. Another example is provided by the motorcyclists whose approach is often to block streets *en masse*, to demand for workers' rights to urban spaces. For informal construction workers, the disruption potential is limited to specific construction sites. In Kenya, a few cases were reported where associations had used disruptive power to storm work sites to be heard as they demanded their rights, in particular, regarding payment owed; yet, these were isolated cases with the ability to disrupt limited to particular building sites. For informal micro-traders, disruptive power is of less potential save from the latent ability to physically block certain areas via demonstrations.

Regime-disruptive power – the ability "to cause disruption to, and on that basis extract concessions from, political elites interested in regime-survival" (Marslev, 2019, p. 18) – is a potentially potent source of power for informal economy workers as an overall group due to their vast numbers and strategic importance, especially as 'vote-banks' during election times. In particular, the fast-growing group of bodaboda riders has become increasingly important in political campaigns, for example, in Tanzania, where the opposition leader would pose with bodaboda riders in the run up to the local elections in 2019. Bodaboda riders are also used strategically in the protests of other groups as, in Kenya, it is now common practice to hire bodabodas to boost street protests regardless of the topic of protest.

However, the common occurrence of splinter-groups and the lack of any consolidated sector wide organizations amongst bodaboda riders and transport workers more generally in both Kenya and Tanzania means that, so far, their potential for exercising regime disruptive power is limited. This notwithstanding, in Kenya, there is a budding national umbrella organization (BAK) of bodaboda riders, which, if well nurtured by members, could provide a platform for negotiation and engagement with authorities. The matatu sector has also occasionally succeeded in mobilizing regime-disruptive power to gain concessions. A similar potential exists in the micro-trade sector due to the vast number of people working in the sector. Nonetheless, even the largest umbrella associations still organize only a fraction of the actual micro-trader populations and hence also in this sector, regime disruptive power so far remain largely as potential.

Societal power

Societal power arises from cooperation with other social groups and society's support for worker demands (coalitional power and discursive power). In the context of informal workers, this relates to coalition making with other informal associations, with formal worker associations, and with broader civil society organizations.

In construction, coalitional power, in general, is absent in Tanzania but evident in Kenya where associations are working with Non-Governmental Organizations (NGOs) and development partners which provide linkages to services and resources. Tanzanian associations are noted to be inward-looking, concentrating on immediate needs without umbrella associations, while, in Kenya, umbrella association exist, although many construction associations and workers are not (yet) members.

Although the micro-trade sector is poorly coordinated, the findings do show potential in greater coordination to enhance associational power as umbrella associations were found to be more robust and able to represent members and facilitate access to services such as loans, training, or health insurance. In addition, a few recent attempts at broader coalitions and cooperation between different types of associations have taken place in Tanzania, perhaps pointing towards a future strengthening of societal power. In addition, support from external organizations such as, for example, foreign NGOs and connections to people in power also figured prominently in some of the cases explored in this book. Hence, in both countries, the two largest and most established umbrella associations (KENASVIT and VIBINDO) had been, and are, receiving support from external organizations such as foreign NGOs and private actors. Finally, favourable statements made by the late president Magufuli in a top-down manner shape public discourses towards a more favourable view on micro-traders which could potentially be harnessed by associations to achieve increased societal power.

Associations in the transport sector, as mentioned, are very fragmented and competitive. Nonetheless, external coordination – like when the International Transport Federation calls for cooperation around campaigns related to the planned implementation of a Bus Rapid Transit system in Kenya – can trigger coalitional power on a case-by-case basis. As mentioned, the nascent affiliation of informal transport associations to established unions is also an example of coalitional power, with potential to spill over into discursive power, like when a group of Bodaboda and COTWU-T marched together on May Day in 2019 in Dar es Salaam with banners that read: "A Boda-boda job is like any other job". The event was reported positively by the media with citizens giving testimony on how much they relied on Bodaboda. This also got the attention of the Dar es Salaam Regional Commissioner and the Mayor of the city, and hence bodaboda representatives were invited to their respective offices for consultation regarding their work.

In summary, as also highlighted in the literature on PRA, we see different power resources reinforcing each other but also observe how most power resources in the informal economy remain underexploited due to fragmentation, lack of resources, and, in general, weak structural bargaining power.

Services provided by associations

Although associations have different frameworks for offering informal social protection, in general, associations remain the bedrock of informal workers' social protection in both countries. While the associations do not use conventional formal social protection concepts such as social assistance, social security, and health insurance, much of the welfare cushioning they provide for workers falls within these social protection categories. It largely falls on preventive social insurance ranging from one-off responses to specific problems to more specific insurance schemes. Almost all associations support members during sickness and death, with a few others intermittently supporting maternity and unemployment. These kinds of support are not comprehensive and vary across sectors, countries, and associations, and some of the main differences emerging across sectors and types of social protection are discussed below.

The following discussion of the specific services offered by associations uses a typology of social protection elaborated by Devereux and Sabates-Wheeler (2004). They distinguish between four main social protection types: protective, preventative, promotive, and transformative. Protective measures largely overlap with social assistance and, as mentioned, are not covered by our project. Preventive measures include both formal social insurance programmes – such as pensions, health insurance, and maternity leave – and informal insurance. Promotive measures function to enhance or stabilize income, consumption, and capabilities and include access to finance or training, while transformative measures address social equity and exclusion, for example, through collective action and representation. As the latter has been dealt with in detail using the PRA, below, we focus on preventive and promotive measures.

Preventive

When it comes to leveraging associations for accessing (preventive) formal social protection in the **construction sector**, both countries do not perform well as indicated by the relatively low share of members that have formal health and pension cover (see Table 10.1). As a result, in both Tanzania and Kenya, construction workers exposed to hazardous working environments have created self-cushioning informal mechanisms, although these schemes only provide partial coverage to workers. In most cases, workers have to pay for themselves since they have neither formal social protection cover nor any collateral and insurance to cushion them from life challenges.

Across the two countries, construction associations ensure that injured colleagues are taken to hospital, with some Tanzania associations supporting colleagues even after being discharged from hospital and while on sick leave. In both countries, employers hardly fully take care of the injured, some make one-off payments. Further, in both countries, some associations had dedicated specific funds to welfare, with one Kenyan association noting that, "we are building our own insurance, having emergency fee for accidents occurring on site". Furthermore, in both countries, construction workers complement group contributions with additional one-off contributions depending on the welfare needs of members at a given time. This is done by mobilizing resources among members as an additional cushioning measure to any member with a critical welfare issue such as admission to hospital or death.

In the **transport sector**, workers are more likely to access formal insurance measures, especially in Kenya due to the SACCOs paying for the workers' NHIF and NSSF coverage. Moreover, for those who are employed by bus companies, the company makes regular contributions to the NHIF to cover the enrolled members. However, as in construction, workers are exposed to occupation health and safety issue and prone to accidents and rely on associations and SACCOs for insurance. The Tanzania associations tend to be stronger in providing informal social insurance with some having different levels, for example, support for death in the immediate family and upon death, survivor's cushioning which is similar to formal survivors' pension, although the social insurance may not be as comprehensive and long term. In cases where associations are not strong, members spontaneously help colleagues who require assistance through informal fund-raising and leveraging of funds. In the Kenyan transport sector, there is more of a mix of formal and informal insurance leveraged through associations. In the latter case, workers' groups sometimes form sub-groups for savings and credit. It seems that many transport workers are reluctant to join formal schemes which partly explain why there are sub-groups to provide informal insurance when SACCOs are encouraging workers to join the formal schemes.

Some associations, in particular, in the **trade sector**, provide for social insurance in their constitutions, specifying the structure of support, rules, and social funds. This structured approach was more latent in Kenya compared to Tanzania, with one association having an emergency fund to which members contribute KES 50 (USD 0.05) daily. In such schemes in both countries across sectors, a member is the point of reference, but other family members, in particular, spouse and children, are largely covered. Other associations extend support to parents and relatives who are part of a members' household. However, support is provided on a reduced level for non-members and mainly includes sickness, in particular, admission to hospital, and death. Other areas which are supported but often on an ad-hoc basis or through loans are school fees, unemployment, funerals, and weddings. As the trade sector is dominated by women, women are, in some cases, also supported during maternity. In Kenya, some associations exempt nursing mothers from attending association meetings and assist women to manage their businesses while on maternity to ensure income for paying association contributions and for survival, while others contribute a given amount to take care of the expenses of nursing mothers. In Tanzania, some associations support women when hospitalized during delivery. Contrary to the Kenya situation, some groups in Tanzania give three months' maternity leave from group responsibilities and subscription fees.

Promotive

Associations in both countries are active in various promotive roles, in particular, supporting members to save and access credit as well as training in sector skills. In the **construction sector** across both countries, the promotive role of associations is very vibrant in terms of providing loans. These benefits are almost equal in both countries with very marginal differences. For example, in Tanzania and Kenya, 20 and 18 percent of members had received loans, respectively. This notwithstanding, the associations in Tanzania which were generally young, were wishing for expanded access to micro-credit, which implies that they have not accumulated enough savings through their associations.

Training through construction associations was more active in Kenya than Tanzania, with associations in Kenya partnering with government agencies and other actors in training, which was offered for free. Furthermore, the Kenya associations were working closely with polytechnics, partnering with them in training graduates in construction skills. These aspects were not manifest in Tanzania although associations had plans to take members for formal training. In isolated cases, some associations in both countries were also supporting members to purchase assets, including land, through loans.

Compared to Kenyan workers, a significant number of **transport** workers in Tanzania rely on associations for access to loans. In Tanzania, SACCOs are primary sources of savings and loans, while for Kenya, there are various types of associations that includes transportation SACCOs mandated by law as well as other SACCOs. In Kenya, associations promote investments through savings, share-holding, or making contributions to purchase SACCO or company vehicles, individual, and association motor vehicles. In Tanzania, associations have encouraged companies to establish lease and buy agreements with members, a few have obtained loans to purchase buses, and for buying motorcycles through loans, while others have approached banks on behalf of their members for loans.

In both countries, associations also facilitate training of members, although Kenya is more vibrant with formal training of members. In Kenya, formal training covers financial literacy, road safety, investment, and exchange visits, while, in Tanzania, informal on-job training was dominant, mainly for motorcyclists, who learn through riding other members' motorcycles.

The **trade sector** is equally active in facilitating members to access loans from associations and private financial institutions. Savings and loans compliment income and cover life contingencies of association members. In both countries, loans are largely from the pooled monthly membership contributions which attract lower rates of interest compared to other financial institutions. Such loans have flexible payment plans, although not infrequently, funds available are not adequate to meet members' demands. In both Tanzania and Kenya, training facilitated by associations focusing on business skills and entrepreneurship was provided by some of the better organized associations. In Tanzania, in some cases, such training resulted in income generating activities and bulk buying for advancement of association members.

Key findings

As discussed in this edited volume, informal workers across different sectors in Kenya and Tanzania share many attributes, including the fact that a significant percentage of workers operate in a fragile environment without adequate social protection. This chapter's comparative analysis of the construction, transport, and trade sectors across the two countries shows many synergies in respect to the working environment and the importance of associations, especially in respect to cushioning workers on welfare issues. Major differences between the sectors relate to the nature of organization, in which the trade and transport sectors manifest a higher level of organization compared to construction sector. The two former sectors have well-established grassroots and umbrella associations, and there are signs of them beginning to join unions for better representation and voice. Apart from these overarching findings, the sector and country comparisons bring out several other key findings as highlighted below.

A higher percentage of workers are enrolled in associations in Kenya than in Tanzania, and Kenya associations tend to overall be more resourceful than Tanzanian associations. In Kenya, workers have higher level of institutional power, especially in the transport and trade sectors, while Tanzania has made some ad-hoc interventions in both sectors which are not institutionalized.

In respect to services, Tanzanian traders are more engaged in social cushioning-type associations, while Kenya traders primarily belong to loan-related associations. In both countries, traders' associations, especially those located in markets, negotiate with local government authorities for trading spaces and infrastructure with different levels of success. In Tanzania, there is nascent trade union involvement with micro-traders, an aspect which has potential for improving the situation of traders. Larger umbrella associations in both countries were found to offer higher levels of support to members, including access to training, infrastructure, and representation. In a way, these umbrella associations play a role similar to unions, and it might be useful for future research to examine this relationship. Although the majority of workers are not in umbrella associations and unions, there are signs, especially in Tanzania, of unions attempting to design appropriate models to accommodate the unique characteristics of informal work.

In the transport sector, there is a remarkable difference between Tanzania and Kenya associations, although, in both countries, workers operate under loose agreement with vehicle owners. Kenya associations operate within a quasi-legal environment – an aspect which should be empowering but it is not, as SACCOs are skewed towards vehicle owners, with minimal benefit to workers. However, well-functioning associations, SACCOS, and companies invest in workers' capacity-building, organize trainings on road safety and financial literacy, and also invest in social protection. In comparison to Kenya, Tanzanian transport workers operate more informally, and the sector is dominated by fragmentation and disintegration of associations, although some Regional Commissioners encourage route associations to form larger umbrella associations for ease of coordination and governance. While this seems to be an advantage to government, associations can exploit such formations to their own advantage in advancing their transformative power.

Construction workers lag behind in organizing in both countries. Tanzania has young associations compared to Kenya, but associational performance levels in both countries are almost similar, except for the ability of Kenya construction associations to use promotive power. They leverage partnerships both at the local and international level to advance the skills of workers.

There are signs of associations exerting different forms of power across the two countries, with associational power being the most vibrant. Through associational power, associations support workers to access preventive, promotive, and minimal transformative services. Structural power is limited in all sectors, although both transport and construction sectors have the potential to disrupt cities and sites of operation to demand rights. In both countries, this power is limited among traders and only used occasionally by construction workers in specific work sites. Societal and institutional power is also present, but more visible in Kenya. In respect to societal power, construction workers in Tanzania are inward-looking, limitedly leveraging support from NGOs, people in power, and private actors. Although the transport sector in both countries is poorly coordinated, on-going external support and coordination has the potential of triggering coalition power for driving the sector needs. On-going external support also features prominently in the trade sector amongst the larger umbrella associations.

In sum, whether in terms of power resources, worker challenges, or associational benefits, most of the differentiation manifests itself on a sectoral basis, in turn, underscoring the importance of adopting a sector-specific lens when analyzing the realities of informal workers.

References

Devereux, S. & Sabates-Wheeler, R. (2004) *Transformative social protection.* IDS Working Paper 232. Brighton, Institute of Development Studies.

ILO (1972) *Employment, incomes and equity: A strategy of increasing productive employment in Kenya.* Geneva, International Labour Office.

ILO (2019) *World employment and social outlook: Trends 2019.* Geneva, International Labour Office.

Jacob, T. & Pedersen, R.H. (2018) *Social protection in an electorally competitive environment (2): The politics of health insurance in Tanzania: ESID Working Paper 110.* Manchester, The Effective States and Inclusive Development (ESID) Research Centre.

KNBS (2019) *Economic survey 2019.* Nairobi, Kenya National Bureau of Statistics.

Marslev, K. (2019) *The political economy of social upgrading. A class-relational analysis of social and economic trajectories of the garment industries of Cambodia and Vietnam.* PhD thesis. Roskilde University.

Riisgaard, L. (2021) *Organizing the informal economy as part of the trade union movement in Kenya and Tanzania*. CAE Working Paper No. 1, 2021. Roskilde, Roskilde Universitet.

ROK (2010) *Constitution of Kenya*. Republic of Kenya. Nairobi, Government Printer.

ROK (2019a) *Basic education statistical booklet*. Nairobi, Ministry of Education, Republic of Kenya.

ROK (2019b) *Kenya population census*. Nairobi, Kenya National Bureau of Statistics (KNBS).

Schmalz, S., Ludwig, C. & Webster, E. (2018) The power resources approach: Developments and challenges. *Global Labour Journal*. 9 (2), 113–134.

Silver, B.J. (2003) *Forces of labor: Workers' movemènts and globalization since 1870*. Cambridge, Cambridge University Press.

Spooner, D. & Manga, E. (2019) *Nairobi bus rapid transit: Labour impact assessment research report*. Manchester, Global Labour Institute.

UN (2019) *World population prospects 2019*. Department of Economics and Social Affairs, United Nations (UN-DESA). New York, UN-DESA.

URT (2019) *Pre-primary, primary, secondary, adult and non-formal education statistics*. Dodoma, Tanzania, President's Office, Regional Administration and Local Government.

11 Concluding reflections

Lone Riisgaard, Winnie Mitullah and Nina Torm

As outlined in the introduction to this edited volume, collective informal bottom-up forms of social protection, while essential in the lives of many informal workers, are notably absent from social protection discussions, and little is known about the extent or the format of these informal social protection mechanisms. In addition, we also noted a gap in analyzing how formal social protection schemes interact with the realities and needs of particular groups of informal workers. With this edited book, we have started to address these gaps.

Recognizing the very heterogeneous and sector-specific nature of the informal economy, we have, throughout the 10 preceding chapters, explored the collective associations that informal workers engage in, the transport, construction, and micro-trade sectors in urban areas of Kenya and Tanzania (in each country, the focus is on the main city and another smaller city). We have analyzed the potential of these associations for providing or enabling access to both formal and informal social protection measures. While Chapter 10 drew out key findings in a comparative perspective related to differences and similarities between countries and sectors, this concluding chapter offers overall reflections on our three specific research questions:

- a Do informal workers, associations offer any kind of informal social protection, and if so, what characterizes the format of these services, who benefits from them, and how do they compare to formal social protection measures?
- b In the formal social protection schemes, how are social protection needs, delivery, and beneficiaries qualified, and to what extent do they cater for informal workers? What, if any, is the role of informal associations in providing access to formal social protection schemes?
- c To which extent is representation of informal workers' viewpoints and realities institutionalized?

In answering these questions, we also highlight our contributions to the literature and discussions around social protection and informal workers. The chapter starts by teasing out our key findings emerging from the comparison between the social protection models conceptualized and implemented 'from above' by public authorities with the models implemented 'from below' by workers' own

DOI: 10.4324/9781003173694-11

collective associations. This is followed by reflections on the power resources available to informal workers' associations and, in particular, their role in filling the representational gap of informal workers – reflections which include assessing the actual and potential role of trade unions in this. Finally, we reflect on how the findings of this book point to a need for expanding the social protection agenda in several ways.

Top-down versus bottom-up forms of social protection

In order to understand and conceptualize the informal bottom-up protection models and how these differ from the formal social protection models, we have analyzed how social protection needs, delivery, and beneficiaries are qualified in the different models. In particular, we have asked of the top-down models, what type of social protection is seen as relevant for informal workers? How are social protection beneficiaries perceived? And what format of social protection delivery is envisioned? Of the bottom-up models, we have analyzed what type of social protection is commonly provided, who has access to it, and what characterizes the format of these services?

Top-down social protection models

Looking at a wider historical context, the perception of social protection can be seen to have travelled from universalism (in Tanzania under Nyerere) to safety nets (in both Kenya and Tanzania under neoliberal restructuring) and, in recent years, towards a more universal discourse. The recent implementation of a universal old age pension in Kenya (and in Zanzibar but not in mainland Tanzania) attests to this. Nonetheless, when it comes to health care (and pensions in mainland Tanzania), current measures implemented and available to informal workers consist of the option to individually contribute to health insurance. Both countries have plans to roll out self-contributory universal minimum health insurance schemes which will be subsidized only for very poor households, identified through means testing (as outlined in Chapter 2).

When it comes to employment-related social protection, systems exist in both countries which guarantee employer-linked pension and health insurance coverage for formal workers, along with other employment-related protections such as maternity leave, severance pay, unemployment benefit, and work-injury compensation. Leaving aside for a moment the issue of health and pensions, our findings strongly suggest the need to rethink the design of employment-related protection. First of all, the current model effectively excludes people working informally from social protection while awaiting the prospect of formalizing their employment relation – something which for the majority is unlikely to occur anytime in the foreseeable future. Second, most people working informally are not in a standard employment relationship with an identifiable employer with whom they could rely on employment-related protection even if formalization was to materialize. This context calls for a conceptual re-thinking of employment-related protection which effectively de-couples it not only from the restrictive formal/informal dichotomy but also from its basis in a standard employment relation which – it is becoming increasingly clear – turned out to be a geographical and historical exception.

With regard to health insurance and old age pensions, there is, as mentioned, an opening towards enabling inclusion of the informal economy through contributory schemes. Our research, however, illustrates how the public schemes seem to not work well for informal workers, as shown in the relatively low enrolment rates. Whereas pension coverage was almost non-existent, health insurance coverage was found to be 41 percent for Kenya compared with 19 percent for Tanzania in accordance with the national averages reported elsewhere (Jacob & Pedersen, 2018; KNBS, 2019). Interestingly, a significant correlation was found between informal association membership and access to formal social insurance, the implications of which we will discuss below. Education levels were also found to be positively correlated with enrolment indicating that coverage is biased towards the better educated. This corresponds well with our findings of a generally low level of knowledge of how formal insurance schemes work, sometimes coupled with beliefs of insurance as bad omen.

Our findings, in addition, show that most informal workers find it difficult to pay the relatively high monthly premiums because of irregular, low incomes. In addition, procedures are considered to be too complicated and the services received inadequate (in relation to the cost). Hence, while formal insurance schemes are now open for informal workers, they remain modelled on the needs and abilities of formal workers and hence require beneficiaries that have the capacity to consistently provide contributions over relatively long-time horizons with a perspective on possible future needs. In other words, the existing formal schemes are not designed to fit the reality of most persons working in the informal economy, indicating the need to redesign schemes. This could include more flexible payment methods and possibly some ability to access unused contributions in the form of loans in order to make the schemes more attractive in addressing short-term needs. In addition, cooperation with workers' own associations also seem to hold potential as discussed below.

Bottom-up social protection models

Although heterogeneity characterizes the informal workers in our research, one aspect that cuts across countries and sectors is the important role played by different workers' associations in filling the substantial formal social insurance gap by cushioning members during difficult times such as sickness, death, and social welfare needs.

While the associations do not use conventional social protection concepts such as social insurance, many of the services they provide for workers fall within social protection if viewed in a comprehensive manner as we have done in this edited book. While what exactly is offered differs widely between groups, it is clear from the data presented and discussed in this volume that informal workers' own associations offer a wider assortment of social protection than public schemes, focusing on preventive and promotive measures and, to a lesser degree, also transformative measures of voice and representation. Associations, however, do not generally offer protective social protection; thus, social assistance in the form of conditional and unconditional cash transfers remains an important prerogative of formal social protection.

Comparing the services delivered by the associations to formal public or private service providers, for example, loans or social insurances, some differences stand out, which speaks to the attractiveness of informal social protection measures, while simultaneously pointing to some of the shortcomings.

Most associations offer preventive measures ranging from ad-hoc contributions when needs arise to more specific insurance schemes, and many also offer promotive measures in the form of savings and loans enabling easy borrowing, but also vocational training, job facilitation, and, to a lesser degree, joint business activities. In addition, some associations play a role in facilitating access to formal services such as loans and health insurance – as mentioned, we found a significant correlation between informal association membership and access to formal social insurance (see Chapter 3 for more detail). Hence, some associations play a crucial role in either registration of members to a formal health insurance scheme or by directly handling members' premium payments via their savings. Although this was by no means common practice, the existence of such facilitation points towards the potential role that some associations might play in encouraging enrolment but also how a group can help prevent default among their members by handling the premium contributions. A particularly interesting example is the KIKOA scheme in Tanzania which was tailored to work through the informal workers' own associations. The scheme attracted considerable interest but alas did not survive, partly due to adverse selection problems (see Chapters 2 and 6).

The informal social protection provided by associations is based on reciprocity and characterized by being personalized, trust-based, and timely so long as a worker is a member and paying dues; while formal insurance schemes procedures for insurance pay-outs can be bureaucratic and time consuming, the micro-insurances were generally described as more timely, flexible, and based on personal trust.

On the downside, cushioning in relation to life contingencies, such as health-related problems, is mainly provided as a one-off pay-out and normally only covers a limited amount. Likewise, while many associations provide some form of savings and loan function which can be essential in smoothening out irregular incomes and addressing life contingencies, they are nonetheless considered insufficient to meet needs. In addition, the limited size and financial resources of most informal associations means that they are vulnerable towards disruptions such as members defaulting or mismanagement of funds by association leaders. Nonetheless, while providing only limited coverage, they do provide social protection services which are for most people in the informal economy difficult or impossible to access elsewhere.

In summary, compared to the top-down formal social protection models, the bottom-up models employed by informal workers' own associations are based on trust and reciprocity and offer flexible but also limited cushioning against more immediate and short-term needs. While this is a better fit with the realities of most informal workers, access to the bottom-up model is often also exclusionary as many well-functioning groups have entry barriers which are likely to take them out of reach for many informal workers. Recalling that, in our study, only 34 percent in Tanzania and 48 percent in Kenya belonged to an association, exclusion, particularly of poorer segments of informal workers, needs to be thought of when considering the possibility of designing social protection measures where access is facilitated through associations. As explored in more detail in Chapter 10, differences between association focus and between members and non-members varied across countries, sectors, and worker-types. However, we did find that in general, and when controlling for other factors, members have higher average earnings compared to non-members suggesting an associational bias towards the slightly better off, although we cannot say with certainty in which direction the causality runs (see Chapter 3 and Torm, 2020 for more detail). Sectoral differences were identified in what was perceived to be the most important benefit of association membership. Hence, controlling for key worker characteristics, trade and transport workers list voice and loans as more important (compared with health benefits), whereas construction workers are relatively more likely to list health benefits as key association functions. Apart from sectoral differences, across sectors and all other things being equal, worker age is correlated with an increased likelihood of listing health benefits over other benefits, whereas women were more commonly own-account and more likely to join associations compared to men. These research findings clearly illustrate the need to consider sector- and worker-specific contexts when discussing social protection. In addition, and in summary, our analysis comparing bottom-up informal social protection with top-down formal models highlight the need to unravel the implicit formal–informal dichotomy and resulting bias inherent not only in formal schemes but also in the social protection agenda and literature more broadly.

The issue of representation

A particular social protection gap identified is that of representation, and again, this is linked to the dichotomy and bias mentioned earlier. Looking at public understandings of, and policies on, social protection, transformative measures, and the issue of representation and voice, in particular, are notably absent. While not framed as social protection but as related to civil rights, Kenya has in, recent years, made moves towards more decentralized and inclusive representation of informal workers, most prominently illustrated in the Micro and Small Enterprises Authority (MSEA) where informal traders are represented amongst other groups. Tanzania does not have such formal institutionalized structures although intermittent support for the informal sector and ad-hoc coordination and consultation does exist.

The overall deficit in formal and institutionalized representational structures relates to the common understanding of informality as a transitionary state on the way to formalization which results in a lack of recognition of people in the informal economy as collective political actors who should have access to represent themselves in their own right in matters which concern them such as in the use of public space or in social protection policies and regulation.

Representation is of absolute key importance if the general wellbeing and resilience of informal workers is to be advanced. As an example illustrated throughout this book, many policies and regulations which govern the workplaces and conditions of informal workers operating in urban space are either ineffective or actually undermine their ability to earn an income and guard their health. Hence, effective representation in matters which affect their work lives is a key necessity. Likewise, it is not just exclusionary but also counter-productive that informal workers – constituting 84 percent and 69 percent of total (non-agricultural) employment in Kenya and Tanzania, respectively (ILO, 2018; KNBS, 2019) – are not directly represented in national-level institutions where issues such as those related to work and social protection are discussed and negotiated. Hence, in terms of representation and voice, the associations which do engage in such activities contribute to filling a large and important social protection gap for informal workers. Institutional linkages and representational capacities have throughout this book been discussed using the PRA, to which we now turn.

The potential of associations – and the power resource approach

Chapter 10 of this book contains a comparative analysis of the power resources available to associations, and hence, here, we focus on the broader implications of our findings. The PRA was developed to analyze the power resources available to trade unions – not informal workers. As such, there are several difficulties in transferring this typology to informal economy workers as we identify below. Below, we go through the concepts of structural and associational power in turn, discussing potential and limitations based on our research findings and including institutional and societal power where relevant.

Structural power

Informal workers, in general, suffer from very low *marketplace bargaining power* as most occupations have low entry barriers, and there is a large pool of available labour. Workers with slightly more but still limited bargaining power include matatu/daladala drivers and skilled construction workers. With regard to *workplace bargaining power*, this was shown to differ greatly by sector as strategic disruption abilities are of key importance for informal transport workers. The potential for transport workers to disrupt operations and hence the circulation of people and goods in the cities is real and acknowledged as such by public officials as a latent threat. This at least partly explains why we see greater efforts at formally regulating this sector – most pronounced so far in Kenya but as discussed with limited results for workers.

Strategic disruption abilities can, if coordinated and scaled, pose a real threat to regime-survival and hence transform into *regime-disruptive power* which, in turn, can be used to "extract concessions from political elites interested in regime-survival" (Marslev, 2019, p. 18). This is potentially a very potent source of power, in particular, for transport workers, but also for informal economy workers as an overall group due to their vast numbers and strategic importance as 'vote-banks' during election times. Currently, as informal worker associations are fragmented in both countries, this remains mostly latent, but the potential is nonetheless real, whether advanced through sector specific coalitions, via cooperation, or even incorporation into established trade unions as discussed below.

Associational power

While associational power, as evidenced throughout this volume, is of immense importance in covering social protection deficits for members, this is nonetheless also where we find the largest conceptual challenges of employing the PRA.

The conceptualization of associational power, as derived from the formation of collective organizations of workers, seems to rest on the assumption that the main purpose of collective organizing is representation – a form of transformative social protection. Nevertheless, as we have shown in the contributions to this book, many informal workers form collectives for other purposes, for example, to find work or to perform preventive and promotive informal bottom-up social protection measures. This was particularly the case with smaller stand-alone grassroots associations which constituted the majority of associations, while it was the larger umbrella associations which tended to engage in representation. We therefore note a general challenge in conceptualizing the capabilities of associations in providing other forms of social protection and hence in extending the PRA framework to grassroots associations.

We note though the considerable latent potential in building stronger and larger networks of associations, particularly with regard to voice and representation, but also to boost infrastructural resources and organizational capacities to leverage preventive and promotive forms of social protection. As of now, even for umbrella associations, representation is most often marginally applied at case by case level and not as broader advocacy directed at general working conditions, at formal social protection policies, or the misfit or lack of enforcement of government rules.

Other factors specific to the informal economy include the multitude of different work and employment constellations – own-account, commission work, wage-worker (without contract), unpaid help, casual work, micro-entrepreneur with a helper, etc. – as well as the often ambiguous nature of employment relations where such exist. In addition, there are often quite distinct power hierarchies between workers within specific sectors such as, for example, between a mason and his helper – something which, in the case of transport workers, has been framed as different 'classes of labour' by Rizzo (2017).

The often unclear nature of employment relations means that there is not always a 'natural' employer to be targeted or engaged with in negotiations, and hence in this book, we left it an open empirical question who formed the relevant object of workers' strategies and possible counterpart in negotiations. Our case studies revealed instances of representation towards employers (in transport and construction), the police (transport and trade), as well as other municipal and national-level authorities (all sectors).

More complicated is the notion of different 'classes of labour' which means that one should not assume automatic solidarity within a collective of workers in a sector, sub-sector, or workplace. This also relates to what we consider to be another challenge of the PRA, namely that it leaves the internal power dynamics of workers' organizations as a black box. In this book, we can therefore be said to suffer from a similar bias as we have focused on social protection services and only to a very limited extent on internal associational power dynamics. Hence, we argue that, while our findings underline the important role played by informal workers' own associations in providing informal social protection, future efforts to rethink formal social protection efforts to work through informal associations would necessitate research on internal associational dynamics which most likely in very real ways restrict access for some members.

The conceptualization of power resources, as developed from the perspective of trade unions, also omits sources of power, which we have found to be of relevance to informal workers' associations. For example, controlling access to space (what could perhaps be called territorial power) which is of key importance to transport workers associations and some micro-trader associations or personalized ties with more powerful actors (a form of societal power) which is important amongst associations in all the sectors studied.

Another strategy to enhance societal power that has been followed by some associations in order to address both resource and representational deficits has been to employ coalition-making in the form of cooperation with external organizations such as foreign NGOs.

Thus, as also highlighted in the literature on PRA, we see different power resources reinforcing each other but we also observe how most power resources in the informal economy remain underexploited due to fragmentation, lack of resources, and, in general, weak structural and institutional bargaining power.

Overall, though associations have in this edited book been viewed as important sources of power which can be leveraged by informal workers to access various social protection services, though with important differences between countries and sectors. Generally, a higher percentage of workers are enrolled in associations in Kenya than in Tanzania, and Kenyan associations tend to be more resourceful. There is a stronger associational life among informal transport workers compared to informal traders and construction workers in both Kenya and Tanzania. In Kenya, workers have some formalized institutional power in the transport and trade sectors, while Tanzania has made some ad-hoc or non-institutionalized rule-bound interventions in these sectors. With regard to addressing the general deficit in institutional power, an innovative strategy that has recently been employed by some associations has been to cooperate with, associate to, or even form new trade unions altogether.

In sum, with regard to the PRA, we have made a conceptual contribution first by identifying some limitations when transferring this typology to informal economy workers. Second, we have shown how the framework, when modified to fit the informal economy and combined with the preventive, promotive, and transformative social protection frame, is useful in highlighting how sector- and country-specific groups of informal workers possess different power resources. Hence it has proved useful in identifying specific challenges as well as existing and potential leverage points, in particular, with regard to issues of transformative social protection in the form of representation and voice.

Tapping into the institutional power of trade unions

The research discussed in this volume reveals that, while most workers belong to associations, very few belong to unions which negotiate, lobby, and advocate for workers' rights. Nonetheless, there has recently been an opening towards informal workers by the dominant union federations and some individual unions. Although still in its infancy this is a significant development. Informal workers can potentially tap into the institutional power of trade unions, while trade unions potentially stand to gain enormously in numerical strength. It is, however, also clear that the recent opening poses a very difficult and politicized challenge for the established unions (for a discussion of this see Riisgaard, 2021). At the same time, not all attempts at cooperation or inclusion of informal workers have been successful as seen in the transport industry where informal workers have largely opted to create new independent unions rather than join established ones.

Most advanced so far has been the transport sectors in both countries where both new and established unions are involved in representing informal workers, albeit in a very fragmented and competitive manner as well as the micro-trade sector in Tanzania where one particular trade union (TUICO) has been active. Below, we discuss the potential of unions for informal workers by comparing them with informal workers' own associations.

Associations or unions – or associations and unions?

Our research provides fertile ground for a comparison between informal workers' own associations and trade unions – an exercise which remains oddly underexplored in existing literature on informal workers (for more detail, see Riisgaard, 2021). Kabeer, Sudarshan, and Millward (2013), based on their study of informal women workers' organizations (primarily incentivized by NGOs) in different countries, argue that collective action in the associations they studied was built around a recognition that livelihood concerns were key and only in addition to this did they focus on a shared identity as workers. The associations we studied were more varied than the ones included by Kabeer, Sudarshan, and Millward (2013) and included associations, particularly, in transport and construction, which were established primarily around occupational concerns. Based on our findings, we would thus come with a more tentative statement recognizing the large variety of associations and argue that, in general, most informal worker associations address broader livelihood concerns often in addition to work-related issues.

Kabeer, Sudarshan, and Millward (2013), furthermore, conclude that a critical lesson deriving from their cases is the need to start with the experiences and realities of the informal workers themselves and that this stands in contrast to the politics of representation traditionally practiced by the trade union movements which generally work through a common set of strategies around a predefined agenda largely based on economic demands and a shared identity as workers.

Looking at trade union engagement with informal workers in Kenya and Tanzania, we do see some trade unions like, for example, KUBHAWO (Kenya Union of Hair and Beauty Workers) in Kenya offering services that go beyond workplace challenges by providing saving and loans services, while other examples include trade unions seeking to engage informal workers via events on general health-related issues such as HIV or Tuberculosis. In general, though, the trade unions tend to focus on work-related issues in their engagement with the informal economy.

It is also clear that even though the unions seek to be innovative in their engagement with informal workers in different ways, they nonetheless follow a more predefined track which is also shaped by the established tripartite institutional frame. This has the advantage of offering access to and experiences with representation – something which is often lacking but greatly needed for informal workers. Unions have national-level institutional representation in different forums including ones related to social protection. On the other hand, unions' experience with representation is largely restricted to negotiations with employers and legal representation of workers via the labour court systems. Hence, if unions are to succeed in transforming into representational structures capable of addressing the representational gap of informal workers, then innovation is needed, particularly with regard to informal workers who do not have an identifiable employer whom to target for negotiations. An interesting approach has been piloted by TUICO (Tanzania Union of Industrial and Commercial Workers) in Tanzania (see Chapter 6) in their attempt to represent micro-traders as they have sought to establish a bargaining committee where different market traders come together at municipal level for negotiation with authorities. Although mixed results were reported by the participating trader associations, this nonetheless provides an example of some of the innovation needed.

Interestingly, we found that trade union engagement in Kenya and Tanzania works primarily through the informal workers' own associations. Hence, we would argue that this does not have to be a question of choosing one organizational form over another but rather an option of choosing trade union representation in addition to the grassroots associations. By implication, this could mean a differentiated set of benefits where unions, for example, provide the transformative measure of institutional representation and lobby for extension of formal social protection schemes and where associations offer the network and social cushioning for short-term needs (preventive and promotive social protection). This, however, would also mean that challenges of entry barriers to informal workers associations are likely to become duplicated as entry barriers to trade union representation.

An important question arises though as to whom unions seek to represent and how that corresponds with the realities and the self-understanding of informal workers? Importantly, many informal workers do not perceive themselves as workers but rather as small businesspeople with small businesses. In addition, some of the services provided by workers' own associations are services one would rather expect from a business association than a trade union – for example, opportunities to network and share information, and resources useful for their businesses activities such as access to credit. Nonetheless, unlike within the trade union movement where inclusion of informal workers is a prominent although highly disputed topic, we did not encounter similar efforts or agendas from the established business associations which cater for formal business (although we did not investigate this specifically). Nonetheless, one could argue for a need to differentiate between people in the informal economy on the basis of their identification (or characteristics) as a worker in an employment relation or as a businessperson with a small business. While there is a need to differentiate, the endeavour is complicated by the multitude of existing work and (often hidden) employment relations as showcased in this edited book. In addition, such differentiations should also be carefully considered, as different representational models have different implications for social protection as discussed below (see also Riisgaard, 2021).

The representational models

As outlined in the introduction, a particular representational model is promoted by the ILO and generally favoured by trade unions. This model posits that existing employers' and workers' organizations should extend membership to workers and economic units in the informal economy, so that they can be represented and participate in social dialogue with a particular focus on transitioning into formality. The social protection implications of this model are largely compatible with the existing formal social protection schemes, where most employment-related social protection will accrue to workers only when they formalize, and where social insurance is accessed through self-contributions.

Another representational model would be to seek to transform the existing tripartite structure to a 3 + 1 model in order to enable informal operators – whether as workers or businesses – to represent themselves on an equal footing with formal enterprises and workers, but as separate from either of these. This is generally the agenda, which is being pushed by transnational advocacy groups like WIEGO (see e.g. Alfers & Moussié, 2019). Within this model, it is questioned whether the vision of full formalization is possible, and emphasis is on approaches which allow for participation of informal workers (in their own right) in policy design and implementation. The Kenyan Micro and Small Enterprises Authority (MSEA) could be seen as a small step in this direction. The implications for social protection of this model are likely to be a more universal and rights-based approach, either through an expansion of 'workers' rights to include all workers or through citizenship claims on the state to guarantee certain welfare rights such as, for example, health care, pensions, and other social security benefits (as has, e.g., been the case with some informal workers' movements in India – see Agarwala, 2013). Amongst the associations participating in our research, such overall claims were largely absent, and focus was on a more immediate level of case by case dealing with challenges faced by members. Nonetheless, we did, in both countries, see examples of informal micro-traders' associations seeking representation alongside business associations and informal transport workers forming their own unions outside the established national federations.

Key implications and further research needs

It is clear from the discussions above that there is no one size fits all solution, and that challenges and needs should be understood in the context of sector and status of work along with associational belonging and other differentiators like gender, age, income, and assets. Nonetheless, in an overall conclusion, we endeavour to tease out the following implications of our research:

There is a need to conceptually re-think and broaden both academic and policy discussions on social protection in order to recognize and address the restrictive formal/informal dichotomy and one-sided focus on formalization as this bias renders most existing formal social protection measures starkly inappropriate and inadequate for the majority of the working populations.

Of particular importance is the representational deficit. Representation should be included in social protection discourses as it is of key importance in terms of ensuring that informal workers have a say in the elaboration of social protection policies and more generally in issues affecting their work and living conditions.

The immense importance of informal workers' own associations in meeting (even if inadequately) the social protection needs of their members needs to not only be recognized but also help inform efforts to reframe national social protection policies and systems.

This edited volume has reported findings from three specific sectors in two specific countries. In other sectors, other sector-specific features will prevail, but the importance of associations in delivering social protection is likely to cut across. While we cannot generalize, it is nonetheless also likely that informal social protection through associations is of great importance in other African countries characterized by a large informal economy and low coverage by formal social protection schemes. On the other hand, we would be cautious in generalizing the specific characteristics of the bottom-up and, in particular, the top-down model of social protection to other contexts, but highly encourage future research in this area. Additionally, our research points towards several other areas in need of further research:

As mentioned, understanding the internal power dynamics of associations and how they might affect members differently would provide much-needed insights to associations' potential as vehicles for representation and for developing stronger coalitions amongst associations as well as in relation to future efforts to enhance informal social protection and possibly link it to formal social protection.

In light of severe disruptions like the ongoing COVID-19 pandemic which has further undermined the formal social protection system and forced reliance on informal cushioning, there is an important need to gain a better understanding of how different types of associations fare during a crisis.

In this edited volume, we have addressed formal social protection measures that have already been implemented, but as also noted, discussions about how to achieve broader inclusion are ongoing, and policies are continually being drafted and pilot schemes launched, as elaborated in Chapter 2. Such innovations and policy proposals need to be assessed for their compatibility with the needs and challenges of particular groups of informal workers. This endeavour should include assessments of innovative solutions implemented in other countries which might serve as inspiration.

Finally, there is a need to further investigate the potential of recent instances of representation of informal workers through formal institutions as demonstrated in the Kenyan case of MSEA and in the case of informal workers' associations joining or forming trade unions. Both hold potential for diminishing the representational gap of informal workers but are hardly explored in existing literature.

References

Agarwala, R. (2013) *Informal labor, formal politics, and dignified discontent in India.* New York, Cambridge University Press.

Alfers, L. & Moussié, R. (2019) Social dialogue towards more inclusive social protection: Informal workers & the struggle for a new social contract. In: *6th Conference of the Regulating for Decent Work Network.* Geneva, International Labour Organization.

ILO (2018) *Social protection for older persons: Policy trends and statistics 2017–19.* International Labour Organization. Geneva, International Labour Office, Social Protection Department. Available from: https://www.ilo.org/secsoc/information-resources/publications-and-tools/policy-papers/WCMS_645692/lang\--en/index.htm.

Jacob, T. & Pedersen, R.H. (2018) *Social protection in an electorally competitive environment (2): The politics of health insurance in Tanzania: ESID Working Paper 110.* Manchester, The Effective States and Inclusive Development (ESID) Research Centre.

Kabeer, N., Sudarshan, R. & Millward, K. (eds.) (2013) *Organizing women workers in the informal economy: Beyond the weapons of the weak.* London, Zed Books.

KNBS (2019) *Economic survey 2019.* Nairobi, Kenya National Bureau of Statistics.

Marslev, K. (2019) *The political economy of social upgrading. A class-relational analysis of social and economic trajectories of the garment industries of Cambodia and Vietnam.* PhD thesis. Roskilde University.

Riisgaard, L. (2021) *Organizing the informal economy as part of the trade union movement in Kenya and Tanzania.* CAE Working Paper No. 1, 2021. Roskilde, Roskilde Universitet.

Rizzo, M. (2017) *Taken for a ride: Grounding neoliberalism, precarious labour, and public transport in an African metropolis.* Oxford, Oxford University Press.

Torm, N. (2020) *Social protection and the role of informal worker associations: A cross-sector analysis of urban sites in Kenya and Tanzania.* CAE Working Paper No. 3, 2020. Roskilde, Roskilde Universitet.

Index

Note: **Bold** page numbers refer to tables and page numbers followed by "n" denote endnotes

Printed in the United States
by Baker & Taylor Publisher Services

Printed in the United States
by Baker & Taylor Publisher Services